NE-Vol. 19

THERMAL SCIENCE OF ADVANCED STEAM GENERATORS/ HEAT EXCHANGERS

presented at
THE 1996 INTERNATIONAL MECHANICAL ENGINEERING CONGRESS AND EXPOSITION
NOVEMBER 17–22, 1996
ATLANTA, GEORGIA

sponsored by
THE NUCLEAR ENGINEERING DIVISION, ASME

edited by
Y. A. HASSAN
TEXAS A&M UNIVERSITY

S. BEUS
BETTIS ATOMIC POWER LABORATORIES

THE AMERICAN SOCIETY OF MECHANICAL ENGINEERS
United Engineering Center / 345 East 47th Street / New York, N.Y. 10017

FOREWORD

In view of ever-advancing heat generator design technology, the ASME Nuclear Engineering Division's Nuclear Heat Exchanger Committee sponsors the two technical sessions on the Thermal Science of Advanced Steam Generators/Heat Exchangers at the 1996 ASME International Mechanical Engineering Congress and Exposition. This volume is a collection of 8 papers presented at the sessions. The sessions bring together experimentalists, theoreticians, and analysts to present their ongoing R&D activities in the area of thermal science. A wide range of technical topics are covered, including heat transfer enhancement under electric fields, thermodynamics of heat pipe systems, investigations of accidents in large experimental test facilities resembling nuclear power plant configurations, simulation of tube bundles using new turbulence techniques, two-phase friction through circular pipes, measurements of cluster-wall contact in fluidized beds, and studies of stratification in reactor components. It is hoped that these papers will contribute to the enhancement of the state of the art in this technology.

The editors would like to express their gratitude to the authors and reviewers for their contributions.

Y. A. Hassan
Texas A&M University

CONTENTS

A STUDY OF BUBBLE BEHAVIOR AND BOILING HEAT TRANSFER ENHANCEMENT UNDER ELECTRIC FIELD

SI-DOEK OH[*] AND HO-YOUNG KWAK
Mechanical Engineering Department
Chung-Ang University
Seoul, 156-756, Korea

ABSTRACT

The effect of d.c electric field on nucleate boiling heat transfer for refrigerants, R11, R113 and FC72 was investigated experimentally in a single-tube shell/tube heat exchanger by using the temperature control method of wall superheat. Also the behavior of bubble under nonuniform electric field produced by wire electrodes was studied by numerical calculation. For R11, the Electrohydrodynamic (EHD) enhancement for boiling heat transfer was observed for all ranges of the wall superheat tested. However, the enhancement in boiling heat transfer disappeared if the wall superheat exceeds 13℃ for R113 and no electric field effect on the boiling heat transfer was observed for FC72. An application of approximately 5kV was enough to eliminate the boiling hysteresis for R11 and R113. Numerical study has revealed that the bubbles are forced away from the heating surface and toward the electrostatic stagnation point by the dielectrophoretic force. Such modified bubble motion turns out to promote the boiling heat transfer if one use proper electrode configuration.

INTRODUCTION

For utilizing low temperature waste heat sources, one of major tasks is to develop a high performance heat exchanger. An especially compact evaporator is an important thermal component for the plants such as Organic Rankine Cycle engine and large scale heat pumps. EHD augmentation[Cooper, 1990] has been proved to be one of the most appropriate technique to enhance nucleate boiling heat transfer in dielectric liquids which are suitable working fluids for the evaporator employed in waste heat recovery plants.

Previous experiments of EHD enhancement in boiling heat transfer have been mainly done on the film boiling regime[Bochirol et al, 1960, Choi, 1962, Markels and Durfee, 1964] where dramatic increase in heat transfer rate occurs. Such great enhancement is known to be due to the film destabilization caused by electrical forces acting on the vapor liquid interface[Johnson, 1968]. However the electric field effect on nucleate boiling has not yet been fully understood. Considerable fundamental research should be done to investigate the nucleation mechanism in cavity under electric field and how dielectrophoretic force due to the difference between the dielectric permittivity of the liquid and vapor phase in nonuniform electric field[Bonjour et al, 1960] affects the bubble behavior near the boiling surface, which in turn promote heat transfer rate. Recently EHD enhancement by up to a factor of ten has been obtained from a lo-fin tube with complete elimination of boiling hysteresis [Cooper, 1990]. The disturbance of heat transfer layer due to the buoyancy driven motion of bubble trapped in weak field region of the lo-fin tube[Cooper, 1990] has been reported to be a cause which bring such dramatic enhancement in nucleate boiling heat transfer. Bubble behavior under electric fields and temperature gradients were studied by Ogata and Yabe[1993].

____

* Now at New Project Development Team
  Hyosung Industries Co. Ltd
  Seoul, 121-020, Korea

One of common findings from the previous investigations on EHD enhancement in nucleate boiling is that the number of bubble increases while the diameter of bubble decreases as the electric field increases[Basu, 1973, Cooper, 1990, Kawahira et al, 1990, Ohadi et al, 1992]. Another interesting observation from previous experiments is the bubble coalescing on the lower part of heat transfer tube surrounded by 6 wire electrodes with equal spacing[Kawahira et al, 1990, Ohadi et al, 1992, Seyed-Yagoobi et al, 1996]. However most of experimental works on EHD enhancement in boiling heat transfer have been done by controlling the heat flux generated electrically. The nucleate boiling heat transfer data obtained by heat flux control, which shows better heat transfer curve, do not provide appropriate data for the design of evaporator utilized in waste heat recovery plants where waste heat rather than electricity is a main source of evaporation of working fluid. Only a systematic experiment on EHD enhancement by temperature controlled method has been reported for the small range of the wall superheat less than 10K[Cooper, 1990].

In this study, the effect of d.c electric field on nucleate boiling heat transfer for refrigerants, R11, R113 and FC72 was investigated experimentally by using a single-tube shell/tube heat exchanger. Broad range of the wall superheat for boiling action was controlled by the temperature of the water flowing inside tube. The wire electrodes employed in this study, which turns out to be appropriate electrode type for tube bundle evaporator are simpler ones so that the bubble behavior under nonuniform electric field can be easily tested and analysed. Also the electric field strength around the heating surface including the electrodes was calculated by solving the Poisson equation numerically to study the bubble motion in nonuniform electric field. Detailed phenomena related to the nucleate boiling under electric field such as effect of the wall superheat, and the forces affecting the bubble motion were investigated in this study.

EXPERIMENTAL APPARATUS AND PROCEDURES

Experimental Apparatus

A schematic diagram of the EHD augmentation boiling heat transfer unit is shown in Fig.1. The experimental unit consists of a single tube shell/tube evaporator(①), condensers(②,③), constant temperature bath circulator(④), hot water storage tank(⑤) and high voltage supplier(⑥). The evaporator shell was made of stainless steel pipe of 150mm inside diameter with two 125mm diameter sight glasses mounted at its midsection to facilitate visual observation of the bubble behavior on the boiling surface.

The test tube in this study was made of brass one with smooth surface. The outside diameter of tube is 19.0mm with 1.2mm wall thickness. The tube length available for heat transfer is about 687mm. Detailed cross sectional view of the test tube is shown in Fig.2. The tube wall temperature were measured at 5 axial stations with equal interval of 150mm as shown in Fig.3. Thermocouple junctions 3, 4, and 5 were for measuring of circumferential temperatures of the tube. Thermocouple junctions used were magnesium oxide insulation T type sheathed with stainless steel tube. Figure 4 shows the method of embedding of the junctions in the test tube wall. The stainless steel sheathed thermocouple of 1.6mm diameter with bare junction was passed into the hole of 1.5mm diameter in the wall and welded at the tube wall to maintain good thermal contact and to prevent leaking of working fluid into the tube. Two additional T type thermocouples were placed at the top and bottom of the shell to monitor the saturation temperature of working fluid as shown in Fig.2. The pressure inside shell was measured by Bourdon type pressure gauge.

Hot water to the tube was supplied from an aluminium water tank. The water was heated by two 1kW and four 100W immersion heaters adjusted by variacs. Fine control of the water temperature in the tank was done by a temperature controller connected to four 100W heaters. A friction pump(⑦) with mechanical seal delivered the hot water from the tank to the tube. Hot water flow rate modified by the by-pass line was determined by means of a calibrated turbine flow meter(⑧ : Kobold Co., DF-48) in the liquid line. A diaphragm type accumulator(⑨) was installed in the liquid line to minimize the fluctuation in the flow rate. The vapor generated in the shell side tube was condensed and returned to the shell by gravity. The cold water supplied from a constant temperature bath (④) circulated the cooling coil of the condenser.

A schematic of the electrode configurations used in this study is shown in Fig. 5. Three electrode configurations were tested. The electrode was made of 0.9mm diameter carbon steel wire coated with copper. As can be seen in Figs.1 and 2, these wires were supported and insulated from the shell and tube by six ring type teflon supporters fitted to the tube. This arrangement with supporters resulted in a constant electric field around the test tube circumference.

A high voltage generator(Kilovolt Co. Model KV 30-20) was utilized to supply high voltage up to 30kV to the electrodes. Also a voltage regulator was used to prevent DC ripples due to the change in the input AC voltage to the generator. The high voltage was fed into the EHD evaporator through a specially modified spark plug fitted in the teflon shell flange. The voltage and current to the electrode were measured by volt meter(Fluke, 80K-40) and multimeter(Barnet Instrument Co., TS 352B/U) respectively.

Experimental Procedures

For each run, degassing of the test liquid was performed by following procedure. After charging of the shell with working fluid, the wire heater winding around the shell was turned on and the hot water in the storage tank was circulated till the temperature of the test liquid reached at the predetermined value, which was continued until the pressure inside shell reaches 3 bar gauge due to the ebullition on the test tube surface. Then cold water was circulated in the condenser coil until the vapor generated in the shell was condensed to reduce the system pressure of 1 bar gauge. Next, the noncondensable gases were purged by opening the valve (⑥). The above procedures were repeated several times for degassing.

The wall superheat of the tube was adjusted by changing the power supply to the immersion heaters installed in the storage tank. Detailed adjustment of the system pressure was done by controlling the cooling water flow rate and the temperature at the inlet of the condenser coil. If the temperature change of the vapor and liquid inside the shell are in the range of $\pm 0.2°C$, the system is assumed to have reached steady state. Once the steady state was established, the temperature and pressure of the system, the temperatures of hot water at inlet and outlet, the temperatures of tube wall and the voltage and current to the electrode were recorded.

For all cases, the boiling experiment at a particular wall superheat without electric field was performed first. Next, the high voltage was applied to the electrode. More detail experimental procedures are shown in Fig.6. The saturation temperature at the experimental conditions were 25.5°C for R11, 48.5°C for R113 and 57.5°C for FC72. The flow rates of hot water, which are very important factor for achieving stable experimental condition were 8l/min for R11 and R113 and 10l/min for FC72. For completion of one set of experiments, ten hours or so were taken. To check the repeatability of the measurements, the heat transfer rate data without electric field were compared with an existing correlation by Stephan and Abdelsalam[1980].

Data Reduction

The heat transfer coefficient, h was determined by

$$h = \frac{\dot{Q}/A}{T_{as} - T_{sat}} \qquad (1)$$

where T_{sat} is the saturation temperature of test liquid and A is the heat transfer area of the tube, which is approximately $0.0371m^2$. The heat transfer rate to the working fluid inside the shell was assumed to be equal to the rate of energy loss of the hot water flowing inside the tube. This is given by

$$\dot{Q} = \dot{m}C_p(T_{wi} - T_{wo}) \qquad (2)$$

where $\dot{m}$ is the mass flow rate of the water inside the tube. The value of the specific heat of water, C_p was taken at the average temperature of bulk inlet and outlet temperature of the water, ie., $(T_{wi} + T_{wo})/2$. The average temperature of the tube wall, T_{as} was obtained by taking of the arithmetic mean of temperatures at the five side wall stations along the tube length with a weighting factor. That Is

$$T_{as} = \frac{(T_1 + T_2 + T_3 + T_4 + T_5)}{5} \cdot TR_m \qquad (3)$$

The weighting factor TR_m is just the ratio of the arithmetic mean temperatures measured at the three circumferential stations to the temperature at side wall, which is given by

$$TR_m = \frac{T_3 + 2T_4 + T_5}{4T_3} \qquad (4)$$

Same circumferential temperature profile was assumed at all cross sections relative to the corresponding tube side wall temperature to obtain this equation.

All T type thermocouples used were calibrated with DC voltage/current standard generator(Yokokawa, 2553) and semiconductor probe. Especially careful calibration was done for the thermocouples to measure the temperatures at the inlet and outlet of the test tube. In this case, the maximum calibrated temperature obtained was -0.3°C at 65°C. The acquisition of data obtained from the T type thermocouples was done by Yokokawa recorder(HR 1310) connected to PC. The data collected from each thermocouple in 2 minutes at 2 second interval were averaged separately to make a data set.

The uncertainties in the temperature and flow rate measurements are estimated to be approximately $\pm$ 1.1% and $\pm 1.5\%$ respectively so that the calculated magnitude of the heat transfer coefficients are accurate within $\pm 7\%$.

CALCULATION OF ELECTRIC FIELD IN A SINGLE MEDIUM

Numerical calculation of the electric field around the heating tube including electrodes in single medium has been done in order to estimate the field strength for dielectric breakdown and the direction of dielectrophoretic force affecting the bubble behavior, in turn, the boiling heat transfer mechanism clearly. The governing equation which is appropriate for the system we tested is 2-D Poisson equation. It is given by

$$\frac{\partial}{\partial x}\left(\varepsilon_x \frac{\partial \phi}{\partial x}\right) + \frac{\partial}{\partial y}\left(\varepsilon_y \frac{\partial \phi}{\partial y}\right) = -\rho_f \qquad (5)$$

with boundary conditions such as

$$\phi = V_1 \quad at \ S_1$$

and

$$\frac{\partial \phi}{\partial n} = 0 \quad at \ S_2$$

where S_1 and S_2 denote for boundaries as shown in Fig.6.

The components of the electric field can be obtained from the electric potential, ϕ.

$$E_x = -\frac{\partial \phi}{\partial x} \qquad (6.1)$$

$$E_y = -\frac{\partial \phi}{\partial y} \qquad (6.2)$$

In this numerical calculation, we assume that there is no free charge ($\rho_f = 0$) and the electric permittivity of medium is constant in the domain considered.

The electrode configurations which we have tested are shown in Fig.5. In the first electrode geometry, the six electrodes locate 5mm away from the tube. The electrodes were oriented at 60° intervals. No electrode was placed just above the top and below the bottom surface of the tube in this configuration. The second electrode configuration can be made by rotating the first one by 30°. The third configuration is just combination of the first and second ones with 8mm away from the heating surface for the electrodes employed in the second configuration.

The computer program utilized in this study is MagNet 5.0 developed by Infoltica Co., Canada in 1994. The program employs FEM for electric field calculation. Grid generation for the first electrode configuration is shown in Fig.7. The number of nodes and triangular elements are 789 and 1452 respectively for this configuration. Same numerical scheme has been used for the second one. More number of nodes and elements have been used for third configuration. Electrical and thermodynamic properties of liquids needed in this calculation are shown in Table 1.

Typical equipotential lines for the first and third electrode configuration are shown in Fig.8. The arrows in this figure show the direction of the electric field. In Table 2, the calculated values of the maximum electric field strength between electrode and the surface of the tube given applied voltage are shown. Thus, the maximum voltage applied to the electrode for dielectric breakdown can be estimated as shown in Table 3, since the dielectric strengths of liquids are known values. However the actual values for the dielectric breakdown have been found to be lower than those calculated one because of the presence of impurities and thermal induced bubbles in liquid[Hara et al, 1987].

The other important result obtained from this numerical calculation is the detailed information on the action of dielectrophoretic force in such a nonuniform field. Assuming that the electric force acting on the bubble is purely dielectrophoretic, the force is given by[Baboi et al, 1968]

Table 1 Electrical and thermodynamic properties of test liquids

Description	liquids	R11	R113	FC72
Average molecular weight		137.37	187.38	340
Typical boiling point (℃)		23.77	47.57	56
Density, 25℃ (g/cc)		1.476	1.456	1.68
Pour point, 1atm (℃)		-111	-35	-90
Kinematic viscosity, 25℃		0.42 Cp	0.66 Cp	0.4 Cs
Critical temperature (℃)		198.0	214.1	178
Critical pressure (atm)		43.2	33.7	18.1
Surface tension (dyn/cm)		22	19	12
Specific heat (cal/g℃)		0.208	0.218	0.25
Heat of vaporization (cal/g)		43.51	35.07	21
Thermal conductivity (Kcal/mhr℃)		0.074	0.057	0.049
Volume resistivity, $\times 10^{11}$ (Ω-m)		1.6414	2.2	0.01
Dielectric strength (kV/mm)		11.6	12.2	14.96
Dielectric constant, 1 atm		2.28	2.41	1.72

Table 2 Predicted values of dielectric strength for different electrode systems at different applied voltages

applied voltage	electrode system 1	electrode system 2	electrode system 3
5 kV	3.3 kV/mm	2.8 kV/mm	2.86 kV/mm
10 kV	6.6 kV/mm	5.6 kV/mm	5.72 kV/mm
15 kV	9.9 kV/mm	8.4 kV/mm	8.53 kV/mm
20 kV	13.2 kV/mm	11.2 kV/mm	11.44 kV/mm

Table 3 Predicted values of dielectric breakdown voltages for different electrode systems for various liquids

liquid \ electrode	first electrode configuration	second electrode configuration	third electrode configuration
R 11	17.57kV	20.71kV	20.27kV
R113	18.48kV	21.78kV	21.32kV
FC72	22.66kV	26.71kV	26.15kV

$$\vec{F}_E = \frac{2\pi D_b^3 \varepsilon_l (\varepsilon_v - \varepsilon_l)}{\varepsilon_v + 2\varepsilon_l} \, \nabla E^2 \qquad (7)$$

where D_b is bubble diameter.

The direction of this force acting on a bubble is always opposite to the direction of ∇E because the permittivity of vapor inside bubble ε_v is less than that of medium ε_l. The direction of dielectrophoretic force on a bubble is given in Fig.9. As clearly can be seen in this Figure, only a bubble which exists in the center lines between the tube and electrodes moves toward the heat transfer surface. On the other hand, bubbles in the other places are forced away from the heat transfer surface and toward the electrostatic stagnation point between electrodes, provided that no other forces are acting on the bubble.

Possible motions of bubbles generated on the heating surface under the buoyancy and electrophoretic forces combined are shown clearly in Fig.10. The bubbles generated on the lower side of the tube move toward and coalesce together near the electrostatic stagnation point where the buoyancy force acting on bubbles is balanced with EHD one. Consequently, the first electrode configuration which provides a electrostatic stagnation point for bubble motion may reduce the heat transfer rate. On the other hand, the bubbles generated on the other places move radially first and quickly rise up due to buoyancy. Similar conclusions for the bubble motion depending on the dieletrophoretic and buoyant forces were obtained in pool boiling under heat flux control[Seyed-Yagoobi et al, 1996] Quite different bubble behavior under the nonuniform electric field generated by the mesh type electrode is plausible[Cooper, 1990]

RESULTS AND DISCUSSIONS

Experiments of nucleate boiling phenomena in nonuniform electric field have been done to investigate the nucleate boiling heat transfer depending on the field strength, the wall superheat as well as electrode configurations employed. Some experimental results obtained are as follows. All the data obtained in this study were taking during the heating up mode.

For R11, boiling heat transfer coefficients as a function of the superheat with different applied voltage are shown in Fig.11. The electrode type employed in this case is the second configuration. The heat transfer coefficients increase steadily as the applied voltage increases for all range of the wall superheat tested. The maximum heat transfer enhancement for R11 is about 230% at the wall superheat of 10.3K as can be seen in Fig.12. This figure also shows that the enhancement magnitude decreases if the wall superheat exceeds 12K regardless of the applied voltage. This may be due to the fact that boiling action by buoyant force is suppressed in the presence of electric field.

The variation of the circumferential temperatures at midsection of the tube for different heat fluxes at zero voltage and the applied voltage of 5kV are shown in Figs. 13 and 14 respectively. As shown in Fig. 13, the temperature drop at tube wall, especially at the bottom of the tube is noticeable as nucleate boiling starts. However such boiling hysteresis is eliminated completely when the moderate voltage of 5kV is applied. It is also noted that once nucleate boiling on the heating surface occurs, the circumferential temperature variation of the test tube appears distinctly, which indicates that the electric field induces and promotes nucleate boiling on the surface. Same circumferential temperature profiles were observed at high voltages with slight decrease in wall temperatures.

Somewhat different results for R113 have been obtained. For R11, steady increase in the heat transfer coefficient depending on the wall superheat has been observed. On the other hand, abrupt increase when electric field is applied and thereafter no appreciable dependence on the wall superheat in the heat transfer coefficients at lower heat flux level have been found for R113. However the boiling heat transfer coefficients with electric field applied is lower than that for zero field if the wall superheat exceeds 13K as shown in Figs.15, 16 and 17. This can be attributed to the fact that boiling action due to the buoyancy -driven force is suppressed by EHD force in this region where many bubbles form on the heating surface. In another words, the paths available for the bubble to depart are restricted[Seyed-Yagoobi et al, 1996] more severly if many bubbles are present. The maximum heat transfer enhancement obtained for R113 is about 280%, as shown in Fig.18. Similarly as for R11, the magnitude of heat transfer enhancement reduces very much as increasing in the wall superheat for R113 with same reason.

In the case of the first electrode configuration, the boiling heat transfer coefficient with the applied voltage of 10kV even decreases when the wall superheat is greater than 10K. This may be due to the fact that the dielectrophoretic force makes bubbles move to the electrostatic stagnation region just below the lower side of the tube so that free movement of the bubbles is hindered very much, which shows clearly how the nonuniform electric field around the boiling surface affects the bubble motion, in turn, the heat transfer mechanism. However the heat transfer coefficient increases again when the buoyancy-driven force dominate over the EHD one at high heat fluxes. On the other hand, the bubbles generated at the lower side of the tube move outward radially and rise by buoyancy for the case of the second electrode, as shown in Fig.10 graphically.

With zero field, the boiling hysteresis was also observed for R113, as clearly seen in Fig.19. Again, moderate applied voltage applied is enough to eliminate the boiling hysteresis for the second electrode configuration as shown in Fig. 20. However no appreciable difference in temperatures at bottom, side and upper surface of the tube, which suggests that the boiling is active for all circumference of the tube at relatively low heat fluxes. It should be noted that the slope of the heat transfer coefficients obtained for R11 and R113 at zero field in this study are lower than those obtained with heat flux control method[Marto and Lepere, 1982, Singh et al, 1993]

As shown in Fig. 21, the boiling heat transfer coefficients for FC72 are insensitive to the applied electric voltages. This can be attributed to the longer charge relaxation time than the bubble departure one so that the bubble behavior is no longer influenced by the electric field[Kawahira et al, 1990]. More detailed discussion on the experimental results for the subject was done by Oh and Kwak[1996].

Flow visualization of nucleate boiling for R113 at the surface heat flux about 30kW/m^2 and several voltage levels is shown in Fig.22. This is the case with employing the third electrode configuration. From the results in Fig.22, it is clear that the applied voltage increase, the number of bubbles increase while the diameter of bubble decreases, which has been observed previously for R11 [Kawahira et al, 1990] and for R123[Ohadi et al, 1992]. At higher voltage, the bubbles are pulled out from the heating surface radially to form a sheet of bubbles between the lower side of the tube and the electrodes. This observation, referred as coalescing effect[Kawahira, 1990], results from the dielectrophoretic force acting on the bubble, as discussed before.

CONCLUSIONS

Experimental and numerical studies were performed to investigate the bubble behavior and nucleate boiling heat transfer with wide range of the wall superheat with temperature control method under nonuniform electric field produced by wire electrodes. The conclusions obtained in this study are as follows

ⅰ) When moderate voltage of 5kV applied, the boiling hysteresis disappeared, which indicates that the electric field applied induces and promotes the bubble nucleation on heating surface although its cause has not been found.

ⅱ) EHD augmentation on nucleate boiling heat transfer is crucially dependent on working fluids. For R11, EHD enhancement in nucleate boiling heat transfer was observed for all ranges of wall superheat tested. However, the enhancement of the boiling heat transfer disappeared if the wall superheat exceeded 13℃ for R113, and no electric field effect on the boiling heat transfer was observed for FC72. The maximum heat transfer enhancement obtained are about 230% for R11 and 280% for R113 at the wall superheat around 10K.

ⅲ) Proper electrode configuration is needed to induce buoyancy-driven bubble motion under electric field, in turn, enhance the boiling heat transfer. Numerical study has revealed that the bubbles are forced away from the heating surface and toward the electrostatic stagnation point produced by the dielectrophoretic force due to nonuniform electricfield generated by wire electrode.

In conclusion, proper electrode configuration, appropriate electric field strength and wall superheat are all important factors to enhance the nucleate boiling heat transfer as well as to prevent reducing the electric field strength for dielectric breakdown, which should be considered for EHD augmented evaporator design and its operation.

ACKNOWLEDGEMENT

The authors wish to acknowledgement the Korea Science and Engineering Foundation for supporting this work under contract number 921-0900-014-2.

REFERENCES

Baboi, N. F., Bologa, M. K., and Klyukanov, A. A., 1968, "Some features of ebullition in electric field," Appl. Electr. Phenom.(USSR), Vol.20, pp.57-70.

Basu, D. K., 1973, "Effect of electric fields on boiling hysteresis in carbon tetrachloride," Int. J. Heat Mass Transfer, Vol.16, pp. 1322-1424.

Bochirol, L., Bonjour, E., and Weil, L., 1960, "Etude de l'action de' champs elestriques sur les transferts de chaleur dans les liquides bouillants," C. R. Hebd, eances Acad. Sci(Paris), pp. 76-78.

Bonjour, E. and Verdier, J., 1960, "Mecanisme de l'ebullition sous champ electrique," C. R. Hebd, Seances Acad. Sci(Paris), pp. 924-7926

Choi, H. Y., 1962, "Electrohydrodynamic boiling heat transfer," Ph. D. Thesis, MIT.

Cooper, P., 1990, "EHD enhancement of nucleate boiling," ASME, J. Heat Transfer, Vol.112, pp. 458-464.

Hara, M., Honda, K., and Kaneko, T., 1987, "D.C. electrical breakdown of saturated liquid helium at 0.1MPa in the presence of thermally induced bubbles," Cryogenics, Vol.27, pp. 567-576.

Johnson, R. L., 1968, "Effect of an electric field on boiling heat transfer," AIAA J. Vol.6, pp.1456-1460.

Kawahira, H., Kubo, Y., Yokoyama, T., and Ogata, J., 1990, "The effect of an electric field on boiling heat transfer of refrigerant-11 ; Boiling on a single tube," IEEE Trans. in Industry Application, Vol. 26, pp. 359-365.

Markels, M. and Durfee, R. L., 1964," The effect of applied voltage on boiling heat transfer," AIChE J., Vol.10, pp.106-110.

Marto, P. J., and Lepere, V. J., 1982, "Pool boiling heat transfer from enhanced surface to dielectric fluids," ASME, J. Heat Transfer, Vol.104, pp. 292-299.

Ogata, J. and Yabe, A., 1993, "Augmentation of boiling heat transfer by utilizing the EHD behavior of boiling bubbles and heat transfer characteristics, Int. J. Heat Mass Transfer, Vol.16, pp.783-791.

Oh, S. and Kwak, H., 1996, "Electrohydrodynamic enhancement on boiling heat transfer of R113+wt4% ethanol," Will be present at the 3rd KSME-JSME Thermal Engineering Conference.

Ohadi, M. M., Papar, R. A., Ng, T. L., Faani, M. A., and Radermacher, R., 1992, "EHD enhancement of shell-side boiling heat transfer coefficients of R-123/oil mixture," ASHRAE Trans. Symp., pp. 427-434, (BA-92-5-1).

Seyed-Yagoobi, J., Geppert, C.A. and Geppert, L.M., 1996, "Electrohydrodynamically enhanced heat transfer in pool boiling," J. Heat Transfer, Vol.118, pp.233-237.

Singh, A., Kumar, A., Dessiatoun, S., Fanni, M. A., Ohadi, M. M., and Ansari, A. I., 1993, "Compound EHD-enhanced pool boiling of R-123 in a liquid-to-refrigerant heat exchanger," ASME Paper #93-WA/HT-40.

Stephan, K. and Abdelsalam, M., 1980, "Heat transfer correlations for natural convective boiling," Int. J. Heat Mass Transfer, Vol.23, pp.73-87.

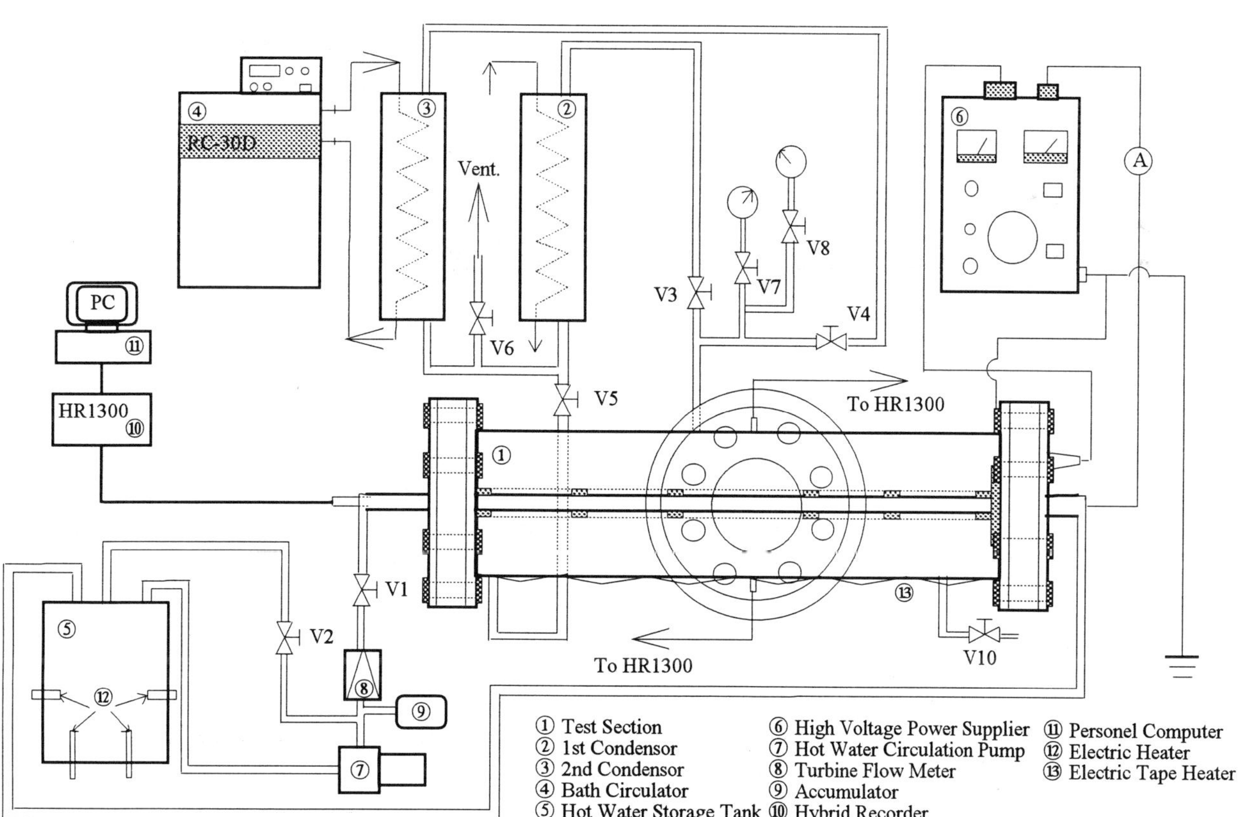

Fig.1 Schematic diagram of experimental loop for boiling experiment

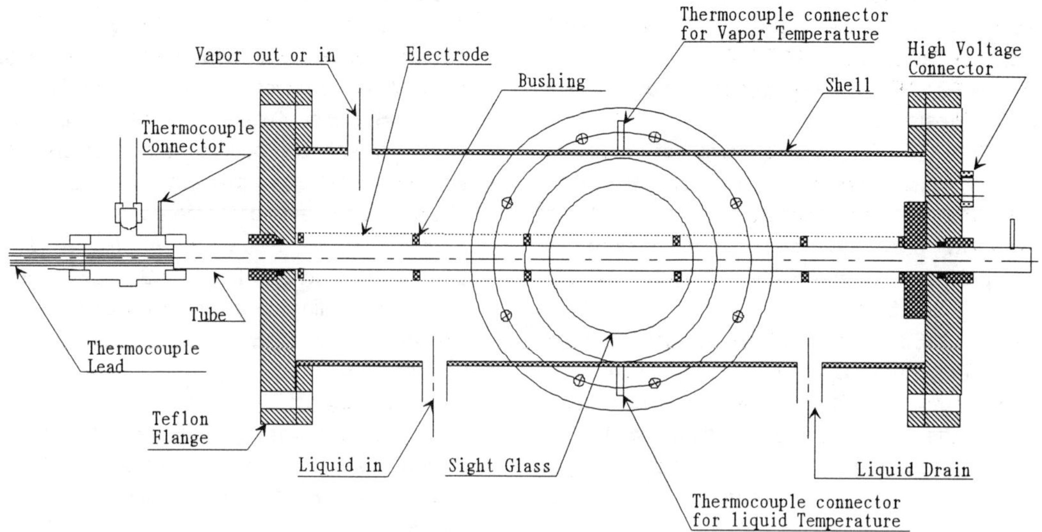

Fig.2 Sectional drawing of the test section

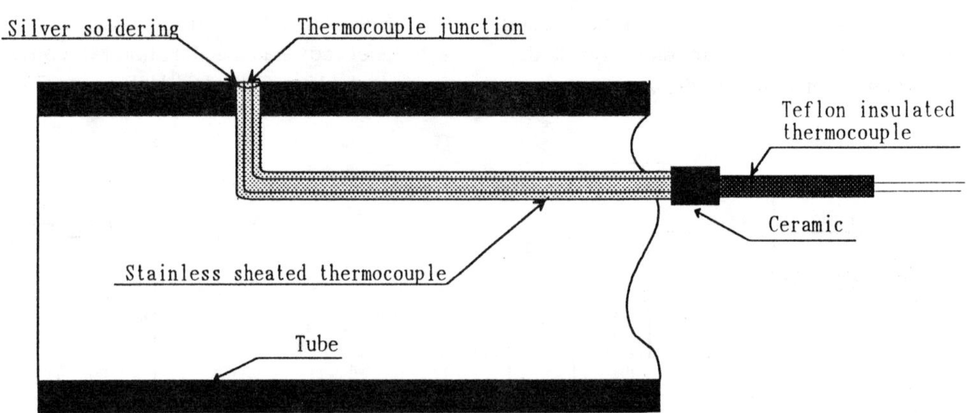

Fig.3 Configuration of the test tube

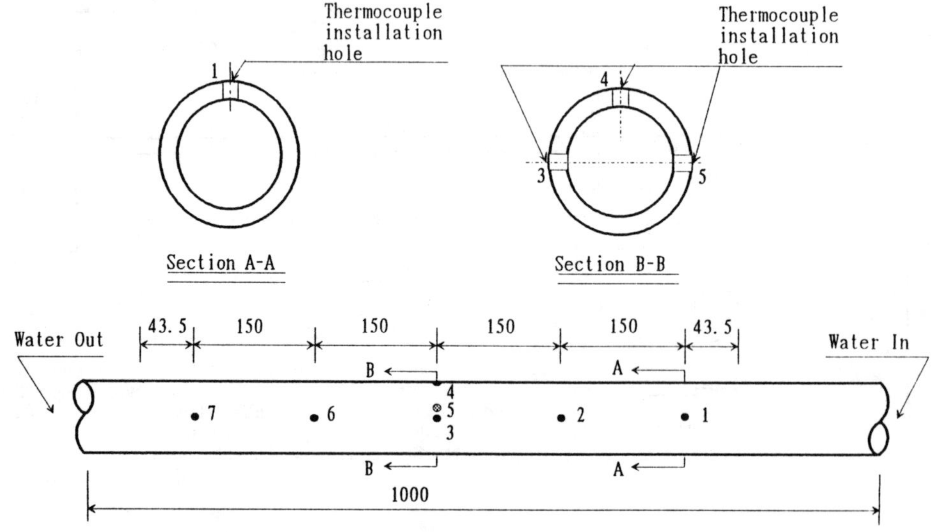

Fig.4 Thermocouple instrumentation of evaporator tube

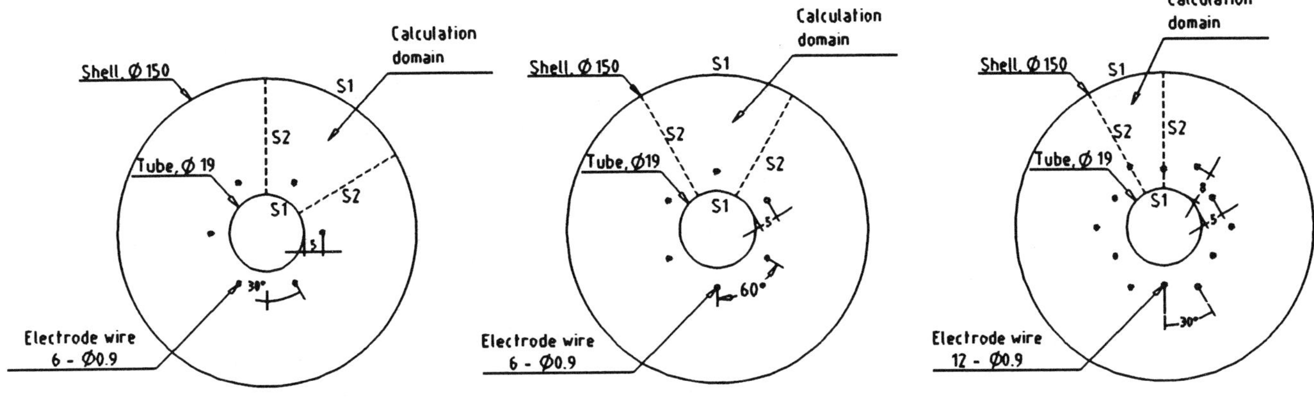

(1) First electrode configuration　**(2) Second electrode configuration**　**(3) Third electrode configuraton**

Fig.5 Electrode configurations for electrostatic analysis and experiments

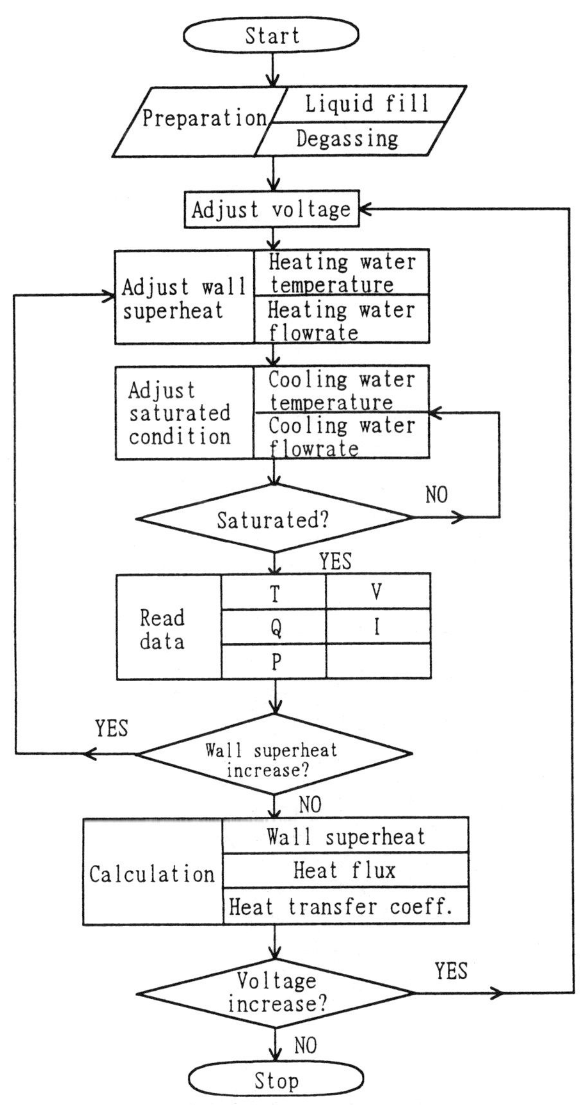

Fig.6 Test procedure of the boiling experiment

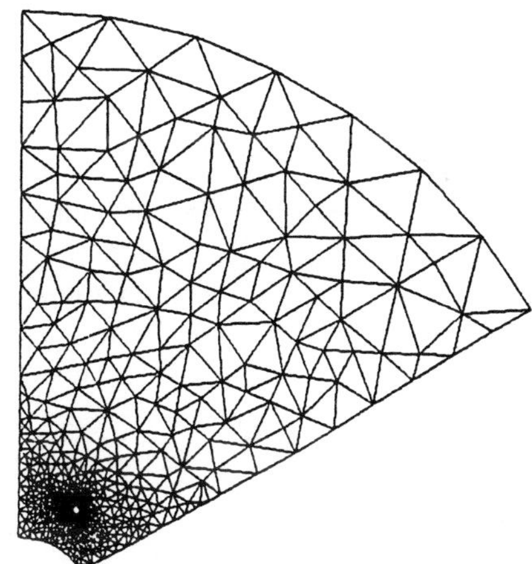

(1) Grid systems of calculation domain

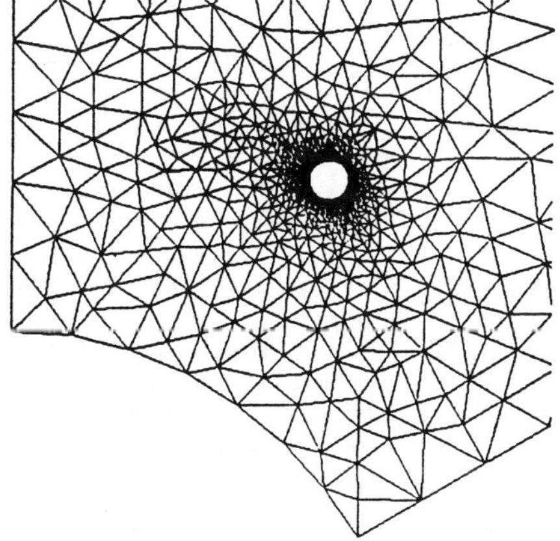

(2) Grid systems around electrode

Fig.7 Grid systems of the first electrode configuration for FEM analysis

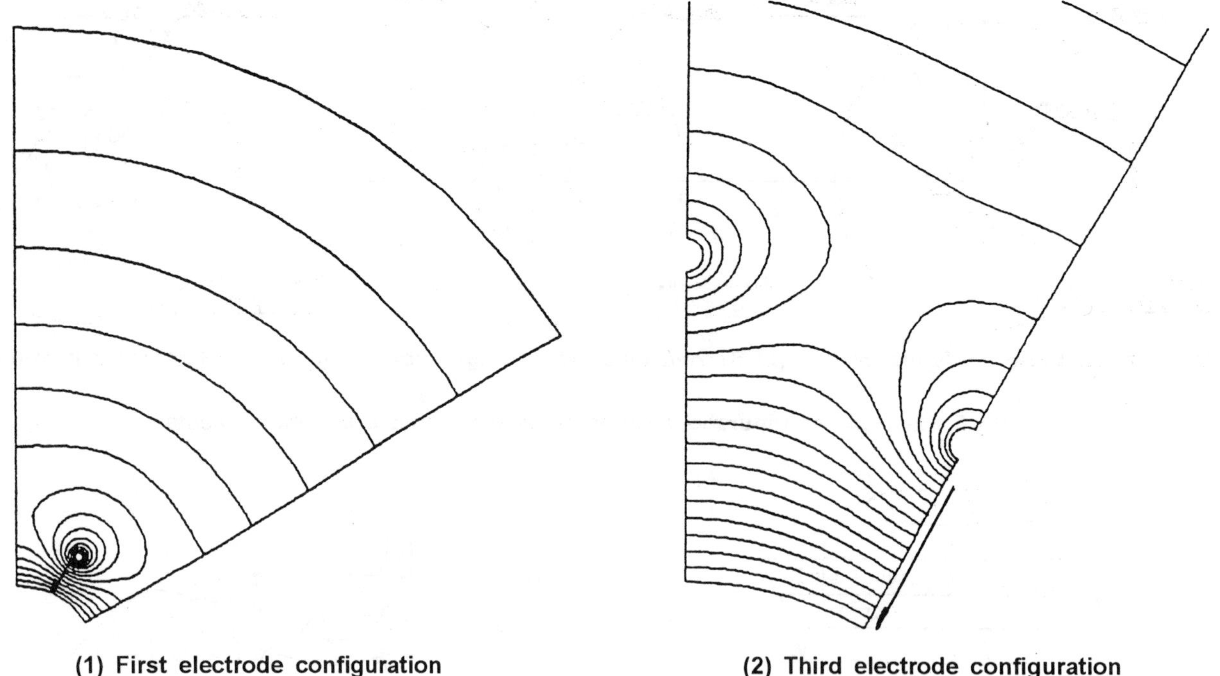

(1) First electrode configuration　　　　**(2) Third electrode configuration**

Fig.8 Equipotential lines for the first and third electrode configuration

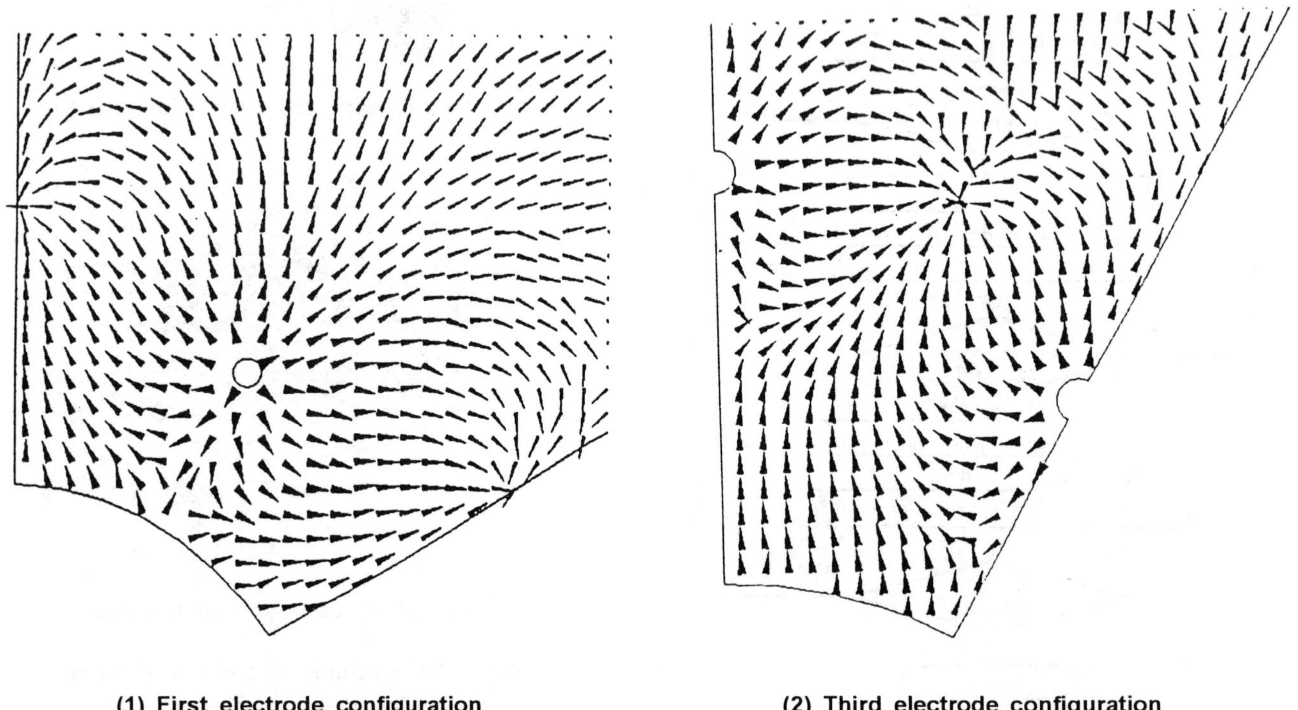

(1) First electrode configuration　　　　**(2) Third electrode configuration**

Fig.9 Directions of dielectrophoretic gradient force ($\overline{F}_E$)

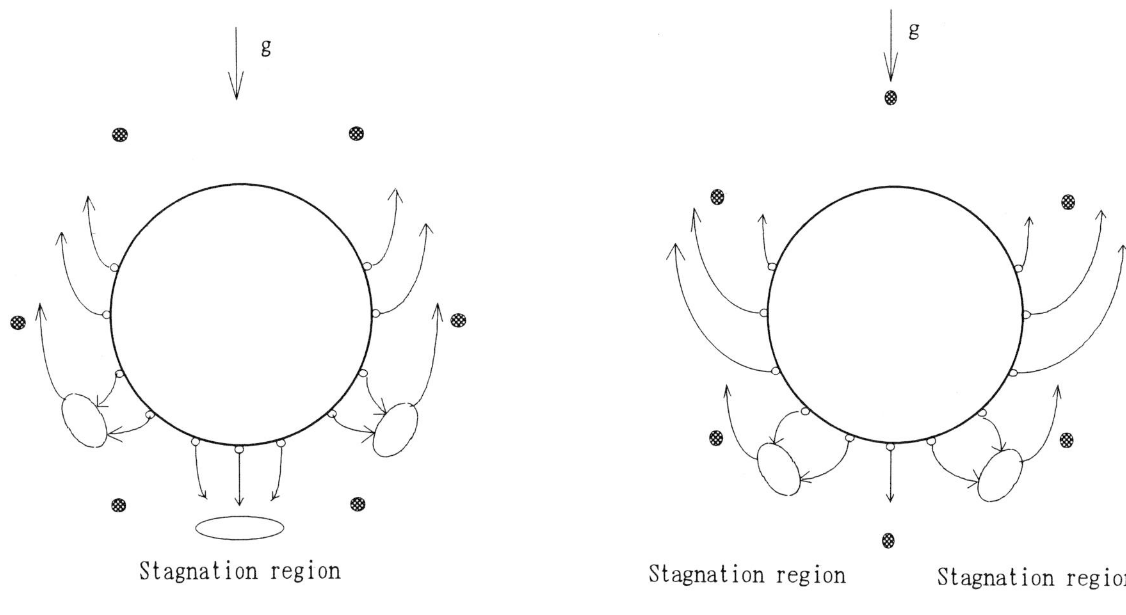

Fig.10 Expected bubble behavior for different electrode configurations

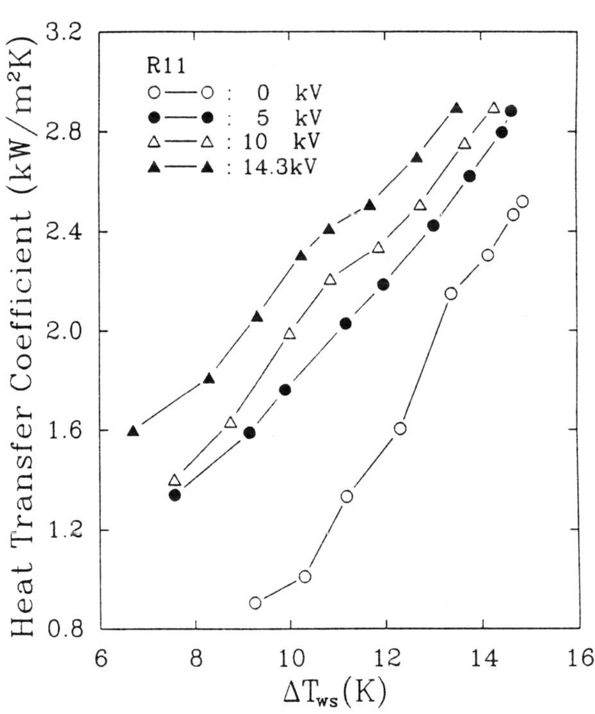

Fig.11 Variation of heat transfer coefficients for R11 with six electrodes (second configuration) and Q_w= 8l/min at T_s = 25.5℃

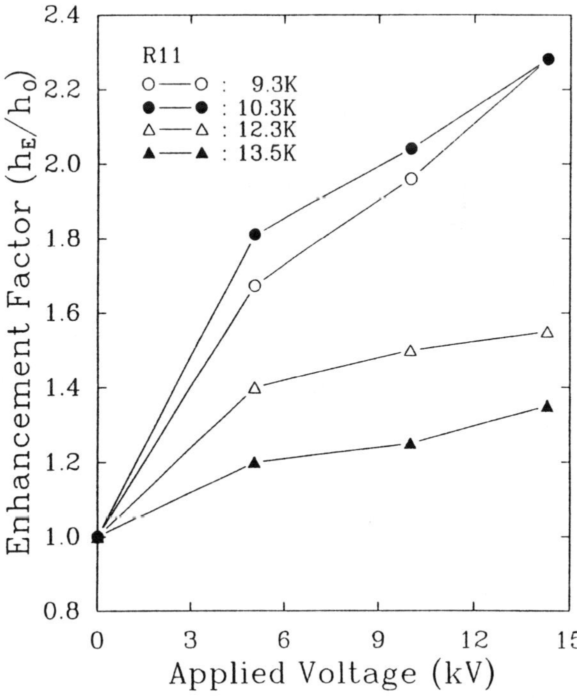

Fig.12 Variation of heat transfer enhancement factors for R11 with six electrodes(second configuration) and Q_w= 8l/min at T_s = 25.5℃

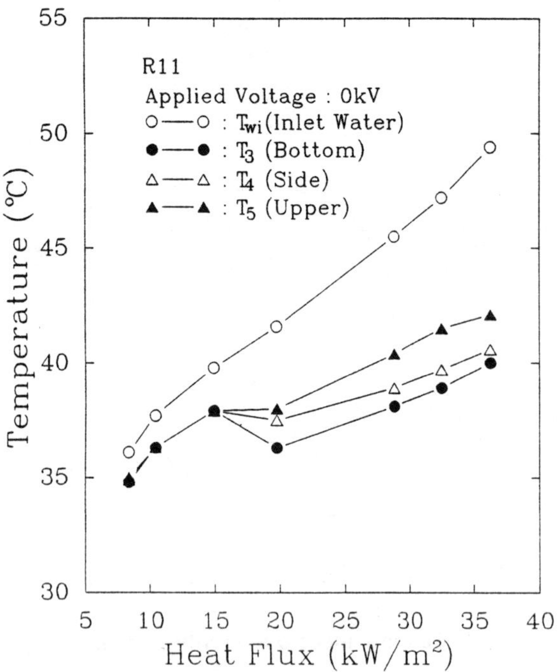

Fig.13 Variation of tube wall and inlet water temperature for different heat fluxes in R11 with six electrodes (second configuration) and Q_w = 8l/min at T_s = 25.5℃

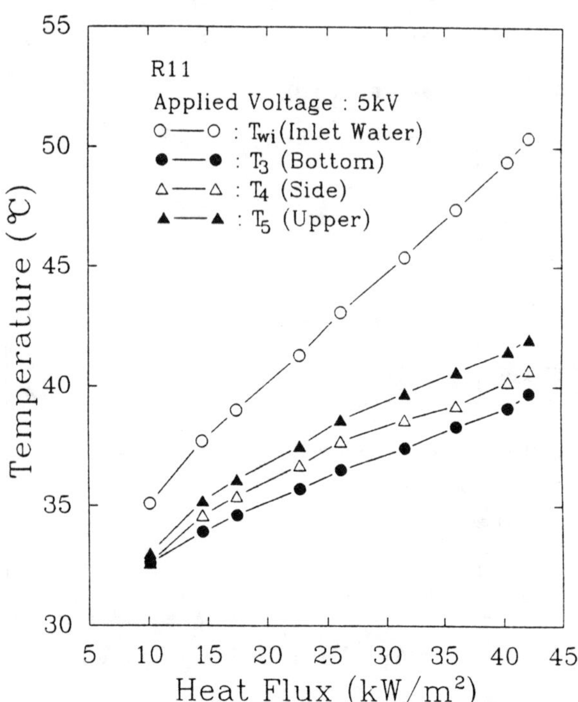

Fig.14 Variation of tube wall and inlet water temperature for different heat fluxes in R11 with six electrodes (second configuration) and Q_w = 8l/min at T_s = 25.5℃

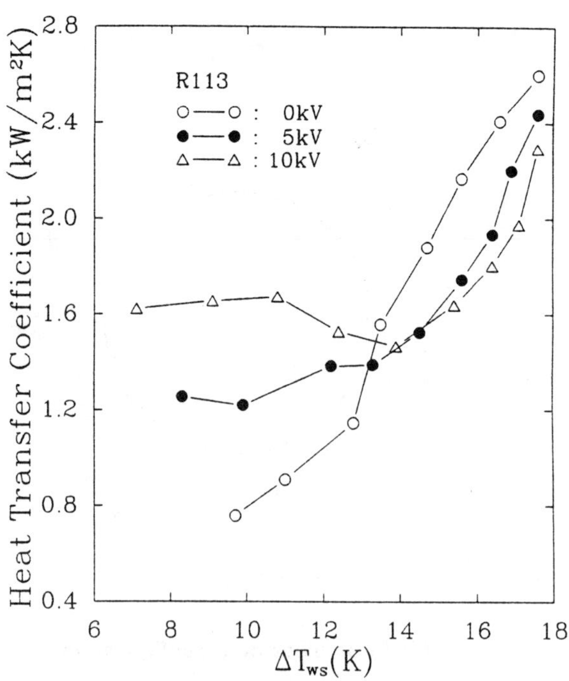

Fig.15 Variation of heat transfer coefficient for R113 with six electrodes (first configuration) and Q_w = 8l/min at T_s = 48.5℃

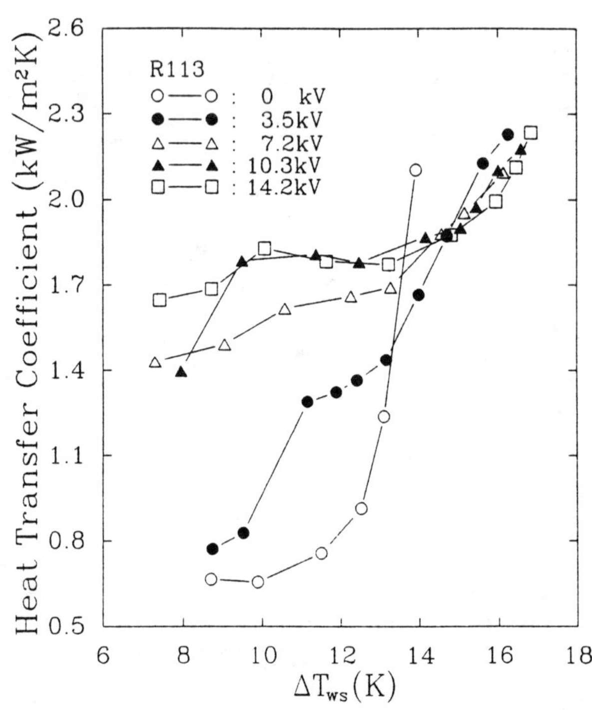

Fig.16 Variation of heat transfer coefficient for R113 with six electrodes (second configuration) and Q_w = 8l/min at T_s = 48.5℃

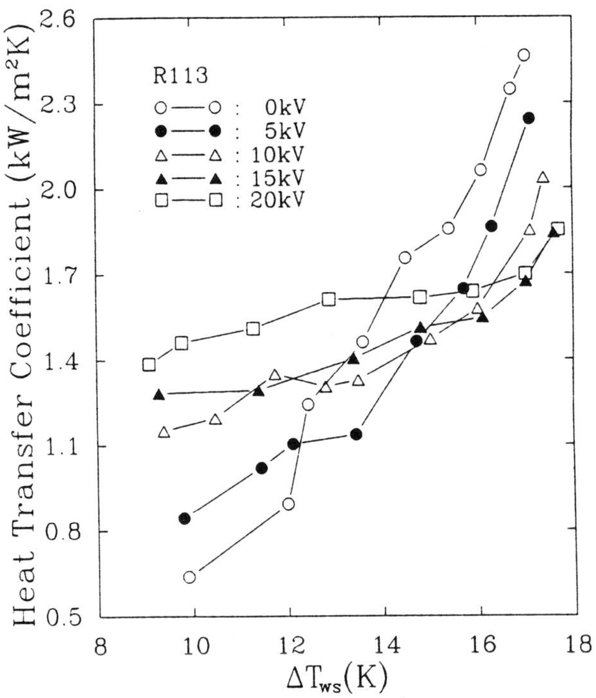

Fig.17 Variation of heat transfer coefficient for R113 with twelve electrodes (third configuration) and Q_w = 8l/min at T_s = 48.5℃

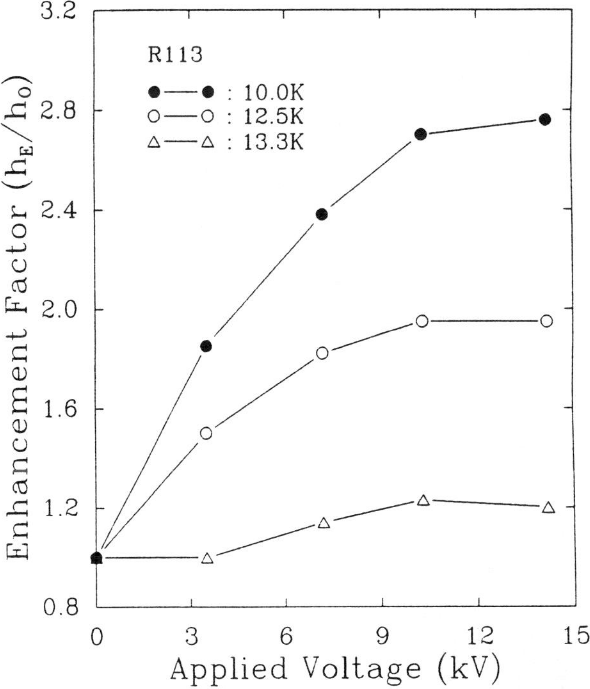

Fig.18 Variation of heat transfer enhancement factor for R113 with six electrodes (second configuration) and Q_w = 8l/min at T_s = 48.5℃

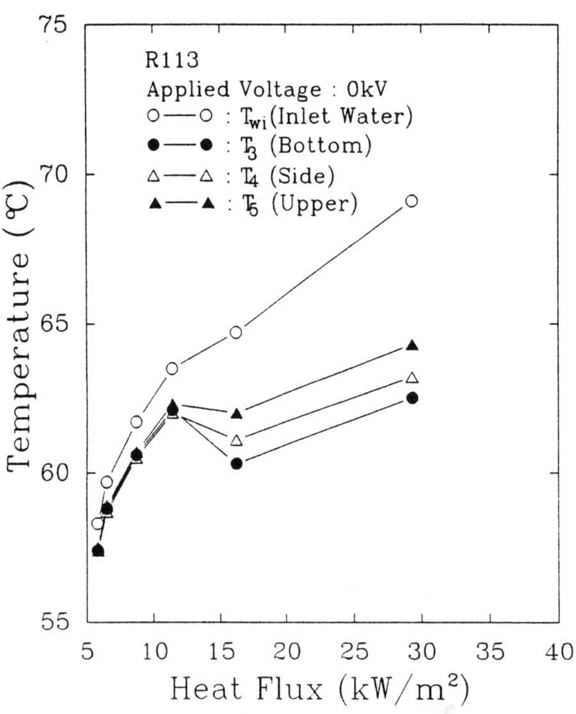

Fig.19 Variation of tube wall and inlet water temperature for different heat fluxes in R113 with six electrodes (second configuration) and Q_w = 8l/min at T_s = 48.5℃

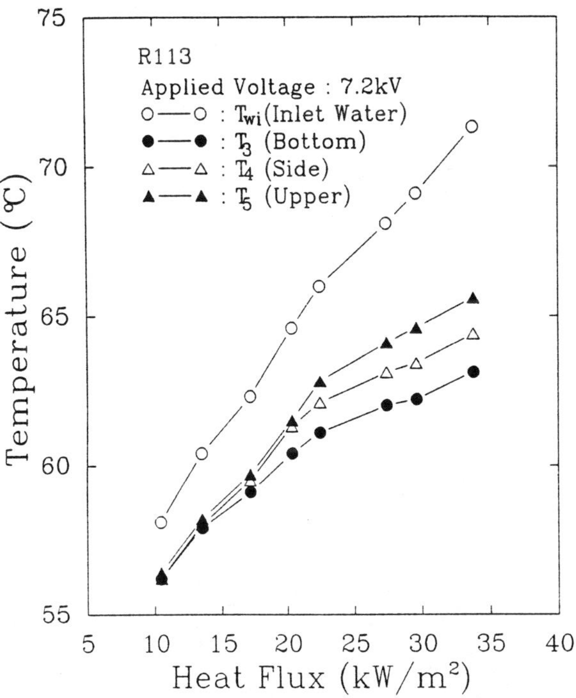

Fig.20 Variation of tube wall and inlet water temperature for different heat fluxes in R113 with six electrodes (second configuration) and Q_w = 8l/min at T_s = 48.5℃

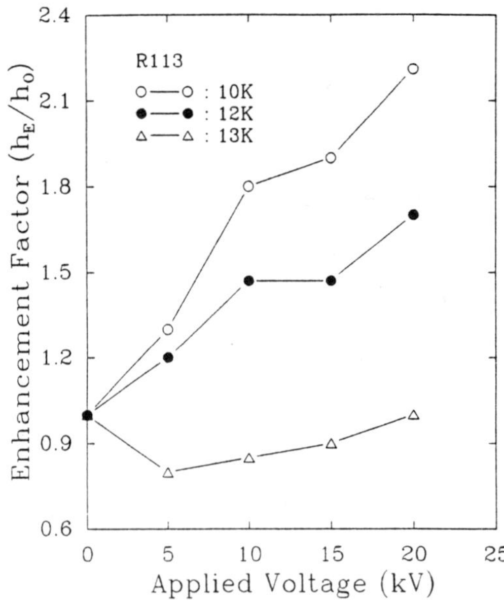

Fig.21 Variation of heat transfer coefficients for FC72 with six electrodes
(second configuration) and Q_w = 10l/min at T_s = 57.5℃

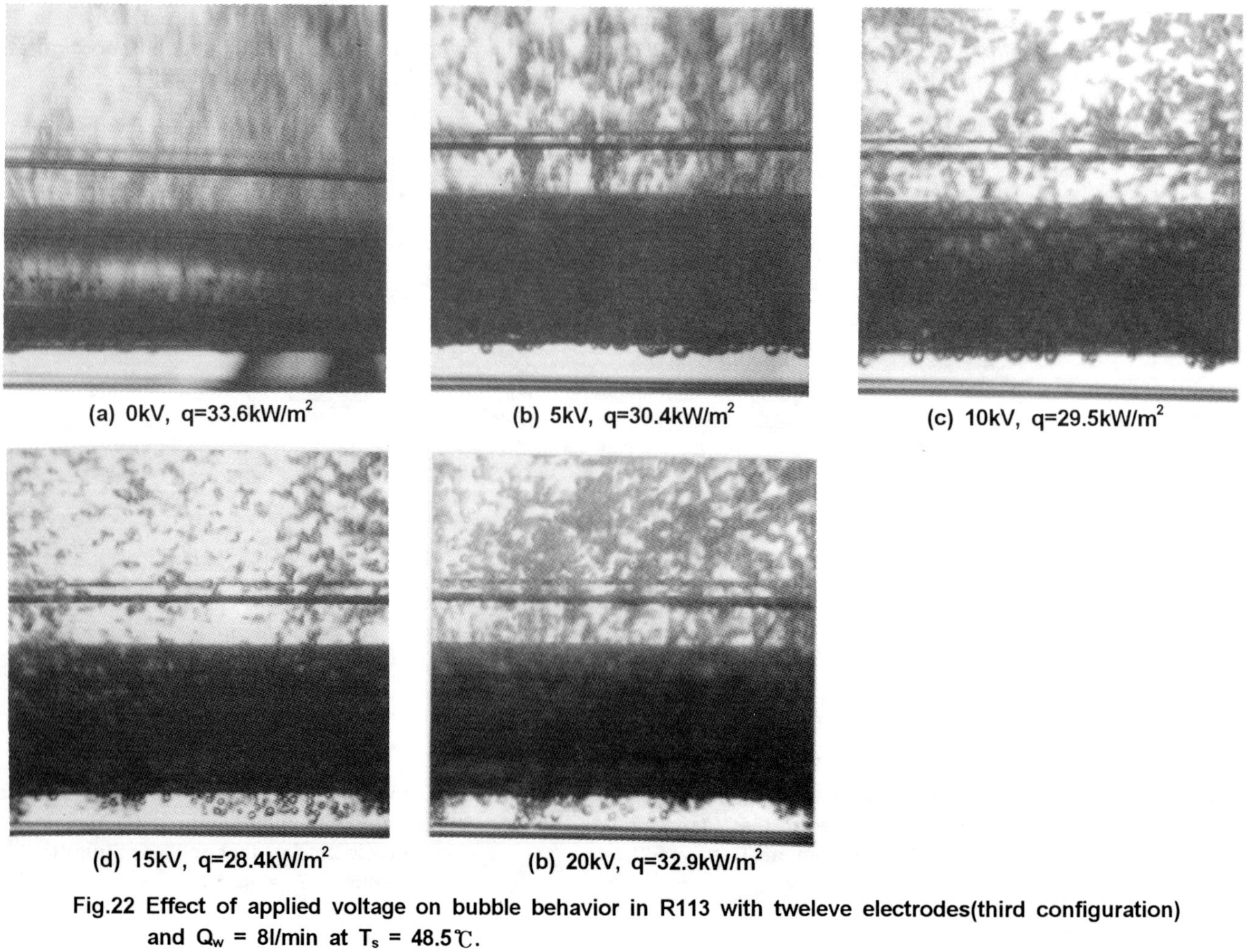

(a) 0kV, q=33.6kW/m^2 (b) 5kV, q=30.4kW/m^2 (c) 10kV, q=29.5kW/m^2

(d) 15kV, q=28.4kW/m^2 (b) 20kV, q=32.9kW/m^2

Fig.22 Effect of applied voltage on bubble behavior in R113 with tweleve electrodes(third configuration)
and Q_w = 8l/min at T_s = 48.5℃.

ENTROPY GENERATION IN HEAT PIPE SYSTEM

Hamed Khalkhali, Amir Faghri and Zhijun Zuo[1]

Department of Mechanical Engineering
University of Connecticut
Storrs, Connecticut 06269, USA

ABSTRACT

A thermodynamic model of conventional cylindrical heat pipes is developed based on the second law of thermodynamics. Entropy generation, an important parameter of heat pipe performance, is caused by the temperature difference between hot and cold reservoirs, the fluid flows, and the vapor temperature/pressure drop along the heat pipe. Ambient temperature in the condenser region and heat transfer coefficient in the transport region can be adjusted to minimize entropy generation in the heat pipe system. A detailed parametric analysis is presented in which the effects of various heat pipe parameters on entropy generation are examined.

NOMENCLATURE

A = cross-sectional area [m^2]
D = diameter [m]
h = heat transfer coefficient [W/m^2-K] or specific enthalpy [J/kg]
h_{fg} = latent heat of vaporization [J/kg]
L = heat pipe length [m]
m = mass flow rate [kg/s]
p = pressure [Pa]
Q = heat transfer rate [W]
S = entropy [J/K]
T = temperature [K or °C]
v = specific volume [m^3/kg]

Greek Symbols

ϕ = ratio of heat transfer coefficients as defined in equation (10)
μ = viscosity [N-s/m^2]
θ = ratio of temperatures as defined in equation (14)
ρ = density [kg/m^3]

Subscripts

∞ = environment
a = transport section
c = condenser
e = evaporator
gen = entropy generation
H = hot reservoir
L = cold reservoir
l = liquid
opt = optimum value
v = vapor
w = wick

Superscripts

$*$ = dimensionless number
$+$ = dimensionless number

---

[1] research associate at the Wright State University, Dayton, Ohio

INTRODUCTION

A heat pipe is a thermodynamic device which is able to transfer heat from one location to another with a very small temperature drop [Faghri, 1995]. Figure 1 illustrates

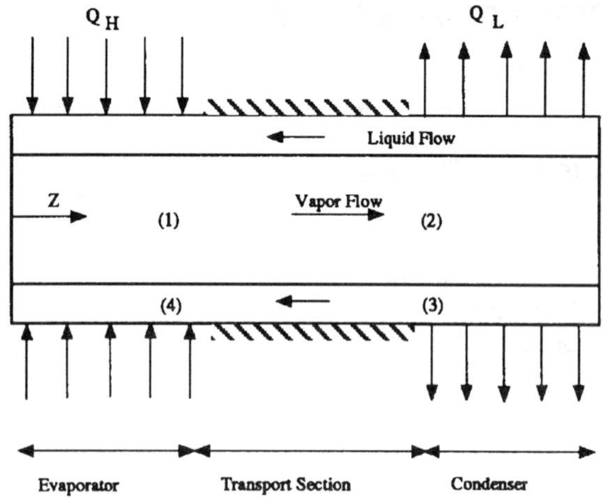

Figure 1 A Sketch of Heat Pipe Operation

a conventional heat pipe with three sections: (1) the evaporator section where heat is added into the system; (2) the condenser section where heat is removed from the system; and (3) the transport section which connects the evaporator and the condenser, serving as a flow channel.

The working fluid inside a heat pipe undergoes a thermodynamic cycle which generates entropy. As in most thermal-fluid systems, entropy generation in a heat pipe is due to fluid flow friction and heat transfer across a finite temperature difference. There exists a direct relation between irreversibility, quantified by the entropy generation, and the amount of lost work during operation. This paper, based on the second law of thermodynamics, seeks to minimize these losses by minimizing the entropy generation. It is shown that minimization of entropy generation can be accomplished by effectively adjusting the heat pipe dimension as well as the external heating and cooling conditions.

Vasiliev and Konev (1990) presented a thermodynamic analysis based on the assumption of constant vapor pressure along the heat pipe. The emphasis of their study was on the evaporation process inside the heat pipe. It was concluded that the overall heat transfer rate can be increased by superheating the vapor in evaporator region. No parametric study of the effects of heat pipe dimensions and external boundary conditions was provided. Additionally, the effect of entropy generation was not discussed.

Richter and Gottschlich (1994) first proposed a thermodynamic cycle analogy to working fluid circulation inside the heat pipe which is illustrated in Figure 2. Pressure variation along the heat pipe was considered. It was assumed that liquid flow in the transport section is adiabatic (i.e. no heat exchange between vapor and liquid or heat loss to the environment). The authors' illustration

of liquid return process (E→A) was rather questionable. Additionally, no quantitative evaluation of heat pipe performance was provided in the paper.

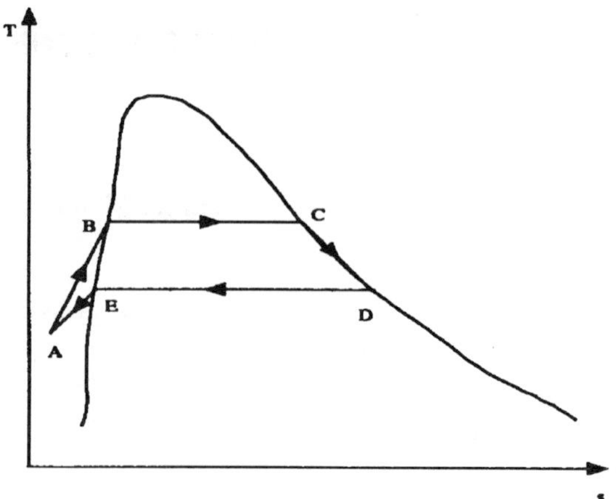

Figure 2 Thermodynamic Analogy of Working Fluid Circulation Proposed by Ritcher and Gottschlich

The scope of this paper is on minimization of entropy generation in the heat pipe. Three factors contribute to entropy generation in the heat pipe: (1) temperature difference between the hot and cold reservoirs (attached to the evaporator and the condenser outer surfaces, respectively); (2) temperature drop in the vapor flow; and (3) friction associated with the vapor and liquid flows. The condenser ambient temperature and the external heat transfer coefficient in the transport section can be adjusted to minimize the entropy generation due both to the temperature difference between the hot reservoir and the vapor and also the temperature difference between the vapor and the cold reservoir. Heat pipe dimensions are the primary adjusting parameters for minimization of entropy generation due to fluid flow friction.

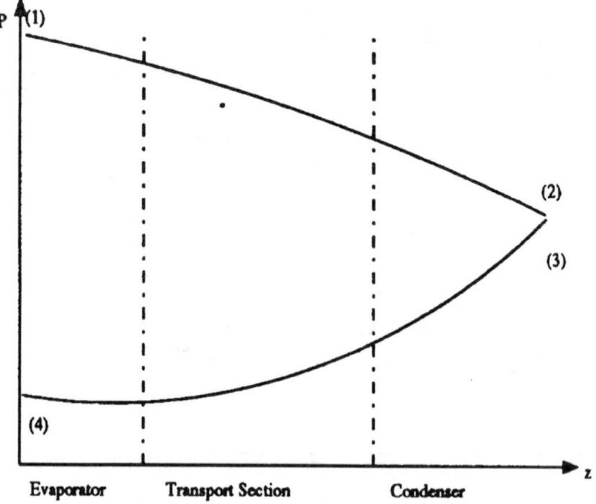

Figure 3 Pressure Variation along a Steady-State Heat Pipe

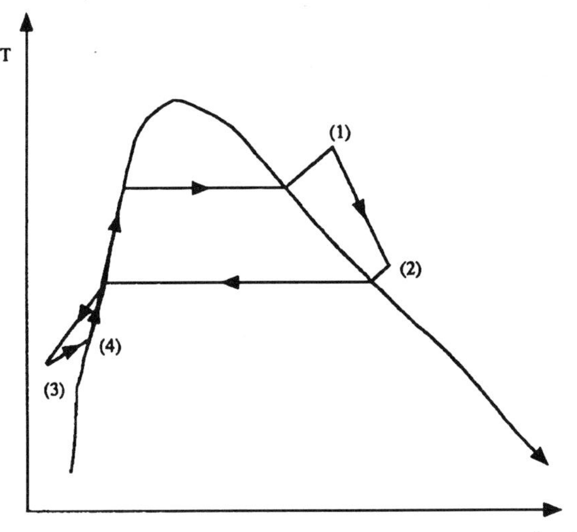

Figure 4   Thermodynamic Cycle of Heat Pipe Working Fluid

Neglecting the effect of gravity, Figure 3 illustrates the pressure variation along the heat pipe at steady state [Faghri, 1995]. States 1 and 2 indicate evaporator vapor and condenser vapor, respectively. States 3 and 4 indicate condenser liquid and evaporator liquid, respectively. It is assumed that the condensation (process 2→3) occurs at a constant pressure. The pressure difference between the vapor and liquid is compensated by capillary pressure which is produced at the curved menisci in the wick.

ANALYSIS

Figure 4 shows the thermodynamic cycle of heat pipe working fluid on a T-s diagram. The following assumptions are incorporated to simplify the analysis: (1) steady-state heat pipe operation; (2) relatively small vapor flow Reynolds number; (3) negligible temperature increase due to viscous dissipation; and (4) no vaporization or condensation in the transport section. The following relationship can be obtained for the working fluid (vapor and liquid phases):

$$dh = Tds + vdp \qquad (1)$$

Assuming an adiabatic transport section, the heat transfer rate in the vapor flow must be equal to that in the liquid flow:

$$(dQ)_{liquid} = (dQ)_{vapor} \qquad (2)$$

Neglecting the effects of kinetic energy and potential energy, application of the first law of thermodynamics to the vapor and liquid flows results in:

$$(dh)_{liquid} = (dh)_{vapor} \qquad (3)$$

Combining equations (1) and (3) provides:

$$(Tds + vdp)_{liquid} = (Tds + vdp)_{vapor} \qquad (4)$$

Since both $(dh)_{vapor}$ and $(dh)_{liquid}$ must be larger or at lease equal to zero and both $(dp)_{vapor}$ and $(dp)_{liquid}$ are negative (pressure drops due to friction), equation (4) indicates that entropy must increase in the vapor and liquid flows. This is illustrated in Figure 4 (processes 1→2 and 3→4). Integration of equation (3) results in:

$$h_1 - h_2 = h_4 - h_3 \qquad (5)$$

States 1 (evaporator vapor) and 3 (condenser liquid) can readily be determined once external heating and cooling conditions are given. So can the working fluid pressure at states 2 and 4. With these parameters known, an iterative procedure involving equation (5) can thus be used to determine both states 2 and 4.

Entropy Generation due to Temperature Differences between Vapor and External Reservoirs

Figure 5 shows a heat pipe attached to a hot and a cold reservoir where the following convective boundary conditions apply:

$$Q_{e,H} = (hA)_H \left(T_H - T_{v,c} \right) \qquad (6)$$

$$Q_{c,L} = (hA)_L \left(T_{v,c} - T_L \right) \qquad (7)$$

$$Q_{a,\infty} = (hA)_a \left(T - T_\infty \right) \qquad (8)$$

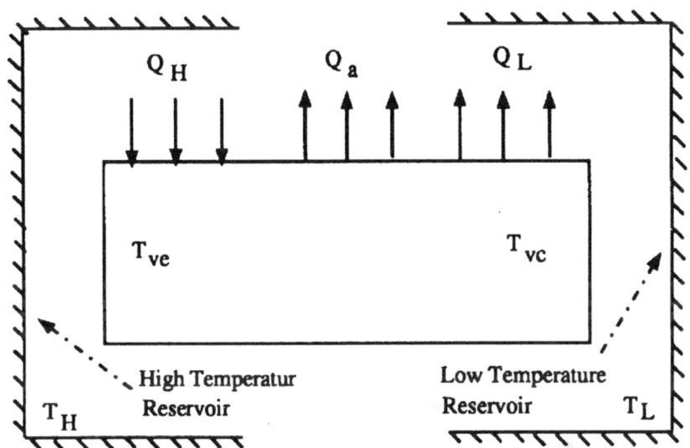

Figure 5   A Heat Pipe Attached to High-Temperature and Low-Temperature Reservoirs

For a low-temperature heat pipe at steady state, the vapor temperature is nearly constant along the heat pipe, i.e. $T_{v,c} = T_{v,e} = T_v$. According to Vasiliev and Konev (1990), the second law of thermodynamics for a heat pipe can be written as follows:

$$S_{gen} = \frac{Q_{c,L}}{T_L} + \frac{Q_{a,\infty}}{T_\infty} - \frac{Q_{e,H}}{T_H} \geq 0 \qquad (9)$$

Substituting equations (6) ~ (8) into equation (9) and using the principle of energy conservation ($Q_{a,\infty} = Q_{e,H} - Q_{c,L}$) yield:

$$\frac{T}{T_H} \geq \frac{\phi \left(1 - T_\infty / T_H \right) + \left(1 - T_L / T_H \right)}{\phi \left(T_H / T_\infty - 1 \right) + \left(T_H / T_L - 1 \right)} \qquad (10)$$

where $\phi = (hA)_a/(hA)_L$. Equation (10) indicates that the vapor temperature T and the external boundary conditions (ϕ, T_H, T_L and T_∞) must satisfy certain correlation according to the second law of thermodynamics. It should be noted that there is no specification of heat pipe material or dimensions in this correlation (equation (10)). Therefore, this thermodynamic requirement is general for any heat pipe.

To minimize the entropy generation, two parameters can be adjusted: (1) condenser ambient temperature T_L; and (2) external heat transfer coefficient in the transport section $(hA)_a$. Based on equation (9) and by incorporating equations (6) ~ (8), an optimum condenser ambient temperature which corresponds to a minimum entropy generation can be calculated as:

$$T_{L,opt} = \sqrt{T_v T_\infty} \tag{11}$$

Applying the principle of energy conservation results in the following two relationships:

$$Q_{a,\infty} = Q_H - Q_L \qquad Q_{a,\infty,opt} = Q_H - Q_{L,opt} \tag{12}$$

Substituting equations (6) ~ (8) into the above equations yields:

$$\frac{(hA)_{a,opt} - (hA)_a}{(hA)_L} = \frac{(T - T_L) - (T - T_{L,opt})}{T - T_\infty} \tag{13}$$

Further substituting the optimum condenser ambient temperature (equation (11)) into equation (13) provides:

$$\phi_{opt} - \phi = \frac{1 - \theta_L / \theta_{L,opt}}{(\theta - 1)/\sqrt{\theta}} \tag{14}$$

where

$$\theta_L = T_L / T_\infty, \quad \theta = T / T_\infty \quad and \quad \theta_{L,opt} = T_{L,opt} / T_\infty.$$

The difference in entropy generation and the minimized entropy generation (obtained by substituting optimum $T_{L,opt}$ into equation (9)) can be calculated as:

$$S_{gen} - S_{gen,opt} = (hA)_L \frac{(\theta - 1)^2}{\theta_L} \phi_{opt}^2 \left(1 - \frac{\phi}{\phi_{opt}}\right)^2 \tag{15}$$

If the transport section is perfectly insulated ($Q_{a,\infty} = 0$ and $Q_{e,H} = Q_{c,L}$), the difference in entropy generation and the minimized entropy generation can be calculated as:

$$S_{gen,ad} - S_{gen,opt} = (hA)_L \frac{\theta - 1}{\theta_L}\left[\phi\left(1 - \sqrt{\theta}\right) + \phi_{opt}^3 \phi(\theta - 1)^2\right] \tag{16}$$

Two dimensionless numbers are introduced to describe the entropy generation:

$$S^* = \frac{S_{gen} - S_{gen,opt}}{S_{gen,ad} - S_{gen,opt}}, \qquad S^+ = \frac{S_{gen} - S_{gen,opt}}{(hA)_L} \tag{17}$$

where S^* indicates the effect of insulation in the transport section, and S^+ indicates the effect of entropy generation minimization. Substituting equations (15) and (16) into the above definitions given by equation (17) results in:

$$S^* = \frac{\left(1 - \phi / \phi_{opt}\right)^2}{\sigma \phi / \phi_{opt} + 1} \tag{18}$$

$$S^+ = \theta_{L,opt}\left(\frac{\theta_L}{\theta_{L,opt}} - \frac{\theta_{L,opt}}{\theta_L}\right)^2 \tag{19}$$

where $\sigma = \dfrac{1}{\phi_{opt}} \dfrac{1 - \sqrt{\theta}}{\theta - 1} - 1$ and $\theta_{L,opt} = \sqrt{\theta}$.

Entropy Generation due to Fluid Flow Friction

The fluid flows inside a heat pipe can be divided into four components: (1) vapor flow in the transport section; (2) liquid flow in the transport section; (3) vapor flow in the evaporator and condenser; and (4) liquid flow in the evaporator and condenser. Entropy generation in any internal flow is due to: (a) flow friction; and (b) heat transfer between the flow and the wall; both of which depends on the mass flow rate or the Reynolds number. As the flow Reynolds number increases, both friction and wall-flow heat transfer increase. An increasing flow friction contributes to a larger entropy generation. On the other hand, an increasing wall-flow heat transfer (correspondingly a smaller temperature difference) tends to reduce the entropy generation. Therefore, there exists an optimum Reynolds number corresponding to a minimum total entropy generation due to the fluid flow.

In the transport section, heat transfer into or from the vapor and liquid flows can be neglected. The main source of entropy generation is the flow friction. As the Reynolds number (or mass flow rate) increases, the friction losses and thus the entropy generation increases.

In the evaporator and condenser sections, both flow friction and heat transfer contribute to entropy generation. As discussed earlier, an optimum Reynolds number thus exists corresponding to minimum entropy generation. As an illustration, the following analysis is conducted on the evaporator liquid flow.

Application of the second law of thermodynamics [Bejan, 1982] to the evaporator liquid flow results in:

$$dS_{gen} = m(s + ds)_l + dm(s)_v - (m + dm)(s)_l - \frac{dQ}{T_{wall}} \tag{20}$$

where $dQ = h_{fg}dm$. Further rearrangement yields:

$$dS_{gen} = s_{fg}dm + mds_l - \frac{h_{fg}}{T_{wall}}dm \tag{21}$$

Assuming the liquid remains saturated in the evaporator region, the following relationship can be obtained:

$$ds_l = -\frac{v}{T}dp_l \tag{22}$$

Substituting equation (22) into equation (21) yields:

$$dS_{gen} = \left(s_{fg} - \frac{h_{fg}}{T_{wall}}\right)dm - \frac{v}{T}mdp_l \tag{23}$$

Further, assuming uniform heat transfer into the liquid flow (which causes liquid vaporization), the liquid mass flow rate in the evaporator linearly decreases in the axial direction:

$$m = M\frac{z}{L_e} \qquad or \qquad dm = \frac{M}{L_e}dz \tag{24}$$

The pressure drop in the evaporator liquid flow can be expressed as follows [Faghri, 1995]:

$$\frac{dp_l}{dz} = -f_l \frac{2m^2}{\rho_l D_{l,h} A_l^2} \tag{25}$$

Substituting equations (24) and (25) into equation (23) yields:

$$dS_{gen} = \left(s_{fg} - \frac{h_{fg}}{T_{wall}} \right) dm + \frac{1}{T} \frac{2 f_l}{\rho_l^2 A_l^2 D_{l,h}} \frac{L_e}{M} m^3 dm \quad (26)$$

where the friction coefficient correlation

$$f_l = \frac{(f\,\text{Re})_l A_l \mu_l}{m D_{l,h}}$$ has been incorporated. Finally,

integration of equation (26) along the axial direction for the entire evaporator region provides:

$$S_{gen} = \left(s_{fg} - \frac{h_{fg}}{T_{wall}} \right) M + \frac{1}{T} \frac{2(f\,\text{Re})_l \mu_l}{\rho_l^2 D_{l,h}^2 A_l} \frac{L_e}{3} M^2 \quad (27)$$

Equation (27) indicates that the entropy generation in the evaporator liquid flow (at a specified heat transfer rate or mass flow rate) is a function of the heat pipe dimensions and working fluid properties.

Entropy Generation due to Temperature Drop along Vapor Flow

For every thermodynamic system, there are two efficiencies defined based on the first and second laws of thermodynamics, respectively. According to Faghri (1995), the first law efficiency of a heat pipe is defined as:

$$\eta_I = \frac{Q_{c,L}}{Q_{e,H}} \quad (28)$$

The second law efficiency of a heat pipe can be expressed as the ratio of inlet to outlet exergies [Vasiliev and Konev, 1990]:

$$\eta_{II} = \frac{e_{c,L}}{e_{e,H}} = \eta_I \left[1 - \frac{1}{(T_{v,e} / \Delta T - 1)(T_{v,e} / T_0 - 1)} \right] \quad (29)$$

where T_0 is a reference temperature usually specified as the condenser ambient temperature and $\Delta T = T_{v,e} - T_{v,c}$.

Entropy generation due to temperature difference between the evaporator vapor and the condenser vapor can be written as:

$$S_{gen} = \frac{Q_{c,L}}{T_{v,c}} - \frac{Q_{e,H}}{T_{v,e}} \quad (30)$$

Substituting equation (28) into equation (30) yields:

$$\frac{T_{v,e} S_{gen}}{Q_{e,H}} = (\eta_I - 1)\frac{1}{1 - \Delta T / T_{v,e}} + \frac{1}{(T_{v,e} / \Delta T - 1)} \quad (31)$$

Equation (31) expresses the dimensionless entropy generation due to the temperature drop along the vapor flow.

PARAMETRIC STUDY

Figure 6 shows the minimum vapor temperature at various cold reservoir temperatures required by the second law of thermodynamics (equation (10)). As expected, the vapor temperature increases with the cold reservoir temperature. In cases that the ambient temperature in the transport section equals the cold or hot reservoir temperature, a linear relationship between the vapor temperature and the cold reservoir temperature is observed. As the ambient temperature in transport section T_∞ decreases, the vapor temperature becomes lower because of

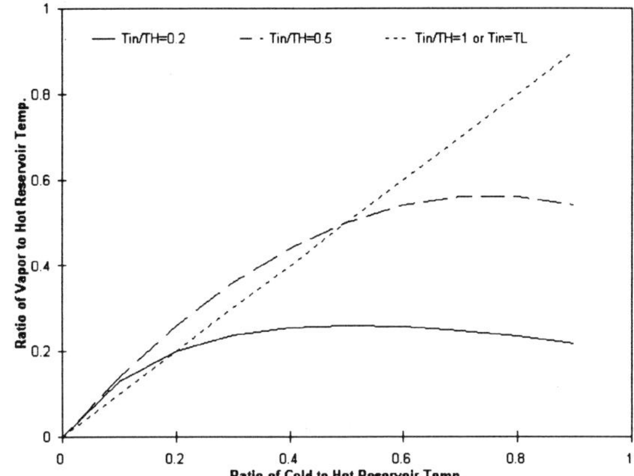

Figure 6 Minimum Vapor Temperature at Various Cold Reservoir Temperatures

the increasing ability of removing heat from the transport section.

Figure 7 shows the minimum vapor temperature at various ambient temperatures in the transport section required by the second law of thermodynamics (equation (10)). As expected, a higher ambient temperature in the transport section corresponds to a higher vapor temperature. Similar trends are observed as shown in Figures 6 and 7.

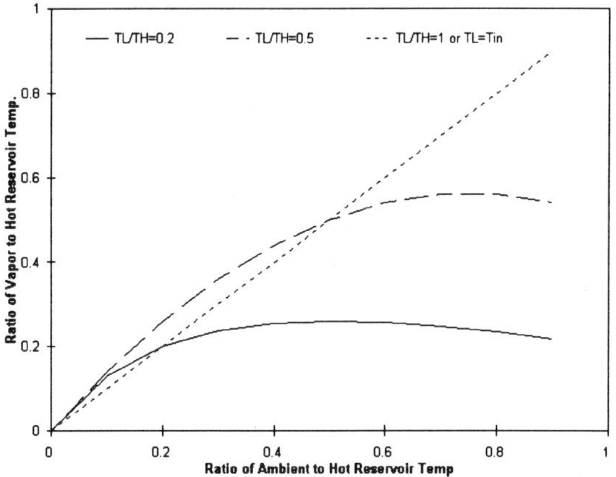

Figure 7 Minimum Vapor Temperature vs Ambient Temperatures in Transport Section

Figure 8 shows the dimensionless optimum condenser ambient temperature ($\theta_{L,opt} = T_{L,opt}/T_\infty$) at various dimensionless vapor temperatures ($\theta = T_v/T_\infty$). As the vapor temperature or heat pipe operating temperature increases, the condenser ambient (or cold reservoir) temperature must correspondingly increase in order to obtain a minimum entropy generation. Figure 8 indicates that: although a large temperature difference between the vapor and the cold reservoir increases the heat transfer rate, it may also increase the entropy generation. Therefore, the condenser ambient (or cold reservoir) temperature should

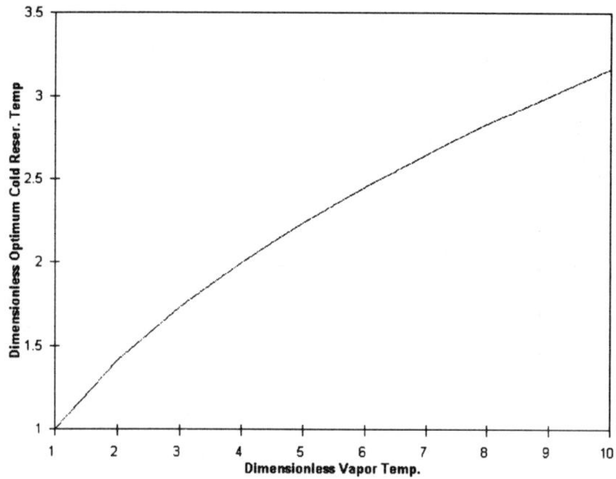

Figure 8   Dimensionless Optimum Condenser Ambient Temperature vs Dimensionless Vapor Temperature

be as close to the optimum value as possible while designing for a heat pipe application.

It can also be seen from Figure 8 that the difference between the vapor temperature and the optimum condenser ambient temperature for higher temperature applications (larger θ) is larger than that for lower temperature applications (smaller θ). This implies that a higher temperature heat pipe can transfer more heat while generating a minimum amount of entropy.

Figure 9 shows the difference between the dimensionless optimum and general heat transfer coefficients in the transport section as a function of the ratio of dimensionless condenser ambient temperatures, which has been expressed in equation (14). It is obvious that the condenser ambient temperature and the heat transfer coefficient in the transport section must simultaneously reach the optimum values (required by the principle of energy conservation).

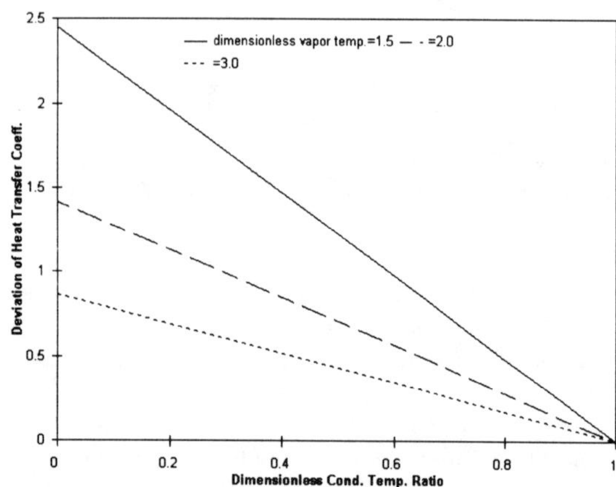

Figure 9   Deviation of Dimensionless Heat Transfer Coefficient vs Dimensionless Condenser Ambient Temperature

Figure 10 shows the dimensionless entropy generation number S^* at various ratios of heat transfer coefficients in the transport section which has been derived in equation (18). When the heat transfer coefficient reaches the optimum value ($\phi/\phi_{opt} = 1$), the entropy generation becomes a minimum value ($S_{gen} = S_{gen,opt}$). When the transport section is perfectly insulated ($\phi = 0$), the dimensionless entropy generation number S^* becomes unity. By increasing the heat transfer coefficient (removing insulation) in the transport section, the entropy generation decreases and reaches a minimum value at an optimum heat transfer coefficient. Further increase of the heat transfer coefficient in the transport section will cause substantial increase in the entropy generation.

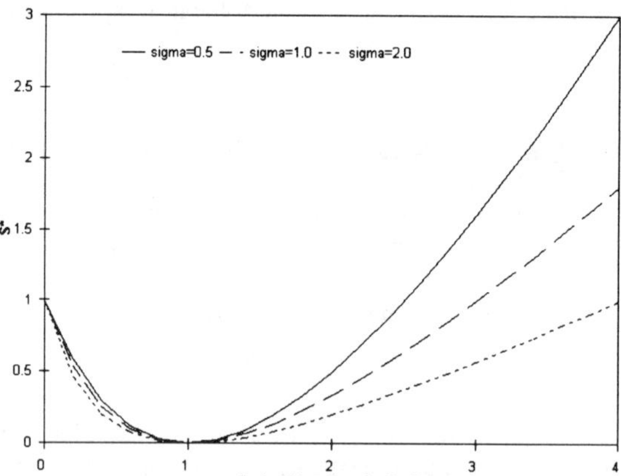

Figure 10   Dimensionless Entropy Generation Number S* vs Ratio of Heat Transfer Coefficients

Figure 11 shows the dimensionless entropy generation number S^+ at various condenser ambient temperatures. As expected, S^+ becomes zero ($S_{gen} = S_{gen,opt}$) when the condenser ambient temperature reaches its optimum value ($\theta_L/\theta_{L,opt} = 1$). As the condenser ambient temperature (θ_L) deviates further from the optimum value ($\theta_{L,opt}$), the dimensionless entropy generation number S^+ becomes larger. This deviation becomes more sensitive at higher condenser ambient temperatures.

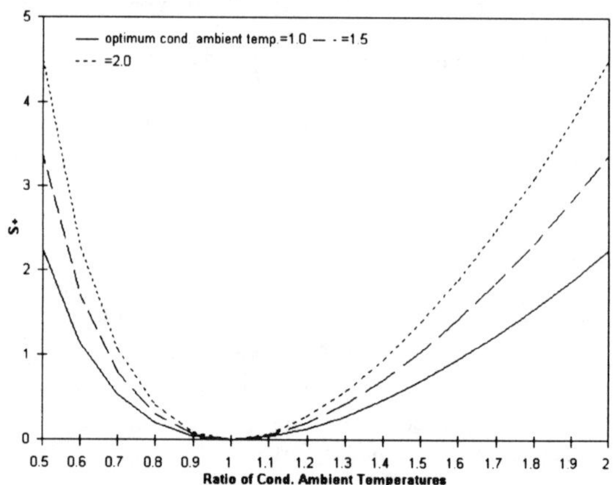

Figure 11   Dimensionless Entropy Generation Number S^+ vs Ratio of Condenser Ambient Temperatures

As shown in equation (27), the entropy generation due to fluid flow monotonously decreases when the hydraulic diameter of the wick increases and the evaporator length decreases. Therefore, while designing heat pipes, the evaporator should be made as short as possible and the wick thickness as large as possible.

Figure 12 shows the second law efficiency of the heat pipe at various values of vapor temperature drop. By assuming a perfectly insulated transport section, the first law efficiency of the heat pipe becomes unity; $\eta_I = 1$. If the vapor temperature remains constant ($\Delta T = 0$), the second law efficiency becomes 100% (no corresponding entropy generation). With the vapor temperature drop increasing, the second law efficiency of the heat pipe quickly decreases. This is even more prominent in the case of low temperature heat pipes (smaller $T_{v,e}$).

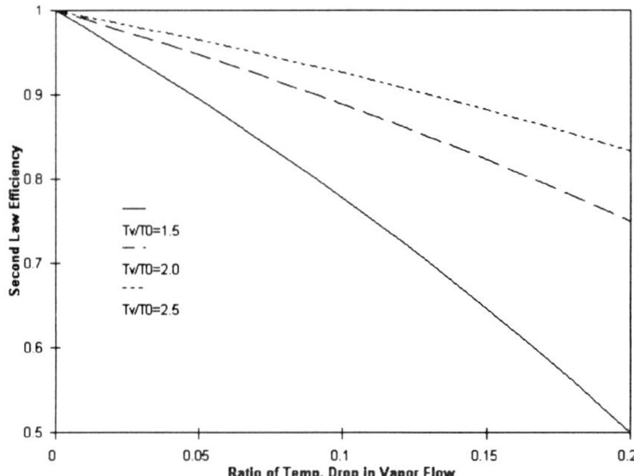

Figure 12 Second Law Efficiency of Heat Pipe vs Temperature Drop in Vapor Flow

Figure 13 shows the entropy generation due to temperature variation in the vapor flow (as expressed in equation (31) for heat pipes with perfectly insulated transport sections). With $\eta_I = 1$, the minimum entropy generation is obtained when there is no temperature drop in the vapor flow.

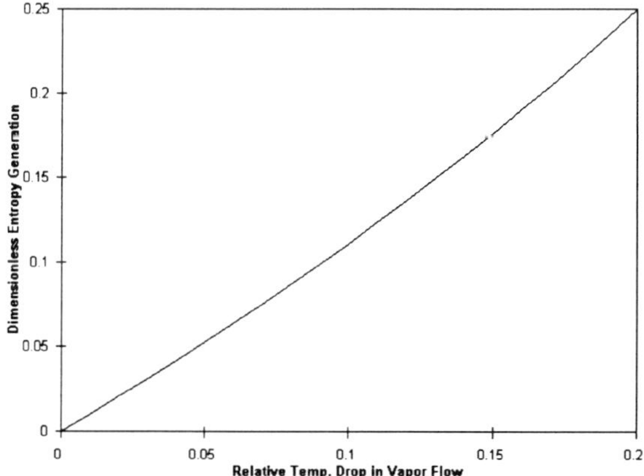

Figure 13 Entropy Generation in Heat Pipe with Perfectly Insulated Transport Section due to Temperature Drop in Vapor Flow

CONCLUSIONS

The following conclusions can be drawn from the above analysis:

(1) there exists a thermodynamic requirement (by the second law of thermodynamics) that the vapor temperature must be greater than a certain value (determined by external heating and cooling conditions) in order for the heat pipe to function properly (see equation (10)). For a heat pipe with a perfectly insulated transport section, this thermodynamic requirement reduces to $T_v > T_L$ which is always true.

(2) there is an optimum condenser ambient temperature which corresponds to a minimum entropy generation. Entropy generation in a heat pipe can also be reduced by removing the insulation in its transport section.

(3) both the condenser ambient temperature and the heat transfer coefficient in transport section must be correspondingly adjusted to obtain a minimum entropy generation.

(4) to reduce the entropy generation due to fluid flow, the evaporator length should be as small as possible and the wick cross-sectional hydraulic diameter as large as possible.

(5) temperature drop in the vapor flow increases the entropy generation and correspondingly decreases the second law efficiency.

An optimum heat pipe design is based on the minimization of entropy generation which ensures minimum loss in the availability in the system. For most cases, heat transfer into the system ($Q_{e,H}$), temperature of the hot reservoir (T_H), and ambient temperature in the transport section (T_∞) are given. The following procedure is recommended to optimize a heat pipe design (minimize entropy generation): (a) choose geometric dimensions of the heat pipe (especially the evaporator and the condenser) to minimize entropy generation due to fluid flows (equation (27)); (b) determine the optimum condenser ambient temperature using equation (11); (c) calculate the corresponding optimum heat transfer coefficient in the transport section using equation (14); and (d) check the performance limitations (capillary, boiling, sonic, ...).

REFERENCES

Bejan, A., 1982, "Entropy Generation through Heat and Fluid Flow", John Wiley & Sons

Faghri, A., 1995, "Heat Pipe Science and Technology", Taylor & Francis, Washington

Richter, R. and Gottschlich, J. M., 1994, "Thermodynamic Aspects of Heat Pipe Operation", Journal of Thermophysics and Heat Transfer, Vol.8, No.2, pp334-340

Vasiliev, L. L. and Konev, S. V., 1990, "Thermodynamic Analysis of Heat Pipe", Proceedings of the 7th International Heat Pipe Conference, Minsk

INVESTIGATION OF BLEED-AND-FEED PROCEDURES FOR MITIGATING THE CONSEQUENCE OF A TOTAL LOSS-OF-FEEDWATER TRANSIENT: IIST EXPERIMENTS

C.H. Lee, T. J. Liu, C.Y. Chang, W.T. Hong, Y.K. Chan, I.M. Hwang, and C.J. Chang

Institute of Nuclear Energy Research

P.O. Box 3-3, Lung-Tan 325, Taiwan, R.O.C.

Fax: 886-3-4711404

ABSTRACT

A reduced-height and reduced-pressure (RHRP) INER (Institute of Nuclear Energy Research) Integral System Test (IIST) facility has been built for simulating the thermal hydraulics (T/H) behavior of Westinghouse three-loop pressurized water reactor (PWR). Experiments on loss of all feedwater (LOFW) with no operator action and with reactor coolant system (RCS) bleed and feed were conducted at the IIST facility. The experimental results not only explore the key (T/H) phenomena during LOFW transients but also show the validation of RCS bleed-and-feed procedures for mitigating the consequence of LOFW transient.

1. INTRODUCTION

The loss of all feedwater transient from PWR power operation conditions will lead to a loss of secondary heat sink followed by a loss reactor coolant system inventory through pressurizer (PZR) power operated relief valves (PORVs). Core uncovery will occur at a RCS pressure equal to or greater than the setpoint of the PZR PORV since the steam generation rate is greater than the venting capacity of the PORVs. The emergency operating procedures (EOP) of bleed and feed for the accident management of the three-loop Westinghouse-type Maanshan nuclear power plant is the process of manually initiating high pressure injection (HPI) and then manually latching open the PZR PORV to depressurize the RCS and allow passage of sufficient water of HPI into cold legs to remove core decay heat for long term core cooling. One of the criteria for initiating RCS bleed and feed is set as any two steam generators (SG) wide range levels less than 6% during the loss of secondary heat sink.

The experimental investigations of the RCS responses associated with accident management for LOFW in a PWR have been recently conducted at several integral system test facilities, such as (1) BETHSY (Chating, 1991) (2) SPES (Billa, 1992), (3) ROSA-IV LSTF (Asaka, 1993) and (4) PKL-III (Umminger, 1995). However, a comparison between two experimental scenarios of LOFW with and without RCS bleed and feed was not known.

With the consideration mentioned above, transient scenarios of LOFW with no operator action and with RCS bleed and feed were investigated by conducting experiments at the IIST facility. The aims of the present paper are to explore the key T/H phenomena during LOFW transients and to assess the validation of RCS bleed-and-feed procedures for mitigating the consequence of a LOFW transient.

2. IIST FACILITY

IIST is a reduced-height, reduced-pressure (RHRP) Integral System Test Facility (Lee, 1991), built to simulate the T/H of the three-loop Westinghouse-type Maanshan PWR under abnormal conditions as well as small break loss-of-coolant accidents (SBLOCA). This facility consists of a core vessel and three loops, each with a steam generator (SG) and a pump. The three loops as shown in Fig. 1 are identical except that one loop is connected to a PZR. The scaling factors of height and volume in the RCS are approximately 1/4 and 1/400, respectively. The scaling of hot leg is based on the Froude number criterion to simulate the transition of flow regimes in the horizontal pipes during accidents. Comparison of the major parameters of IIST and Westinghouse Maanshan PWR are listed in Table 1.

The data acquisition system of IIST involves over 200 instruments located in the primary and secondary sides, used to measure temperature, pressure, flowrate, liquid level and differential pressure. In this facility, 50 view ports are specially designed for flow visualization which lead to significant improvement in the understanding of various two-phase phenomena within the pressure vessel, hot legs, SG inlet and outlet plena, SG secondary sides, crossover legs (loop seals), cold legs, and PZR. A total of 13 video cameras are mounted at selected view ports to record the key T/H phenomena during the experiments.

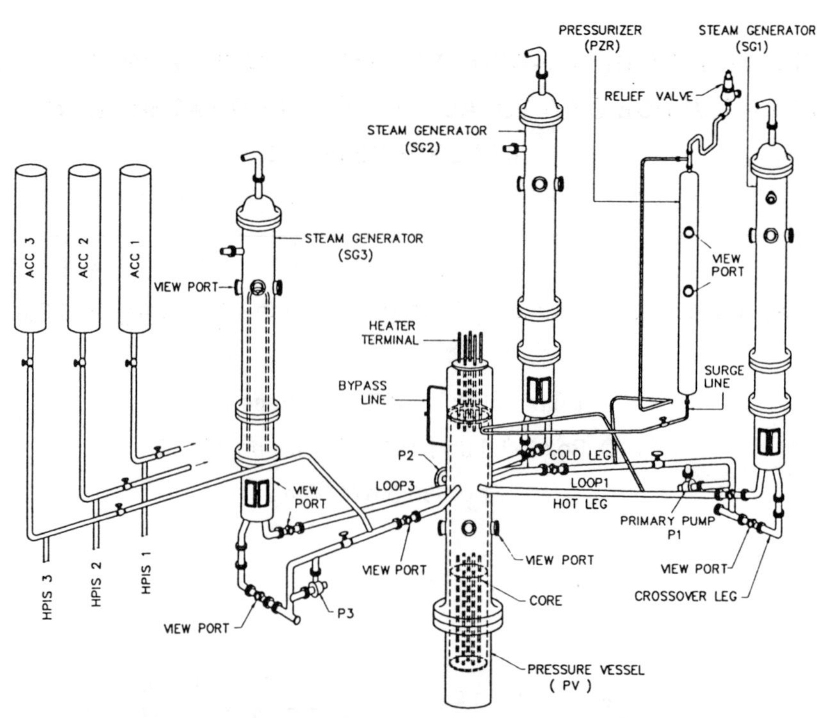

Fig. 1   Schematic of INER integral system test (IIST) facility

Table 1. Comparison of the major parameters of IIST and Maanshan PWR

Parameter		IIST	Maanshan PWR	IIST/PWR
Design pressure	(MPa)	2.1	15.6	1.35×10^{-1}
Max. core power	(MW)	0.45	2775	1.62×10^{-4}
Primary system volume	(m³)	5.37×10^{-1}	2.15×10^{2}	2.50×10^{-3}
No. of loops		3	3	1.0
Core				
Height	(m)	1.0	3.6	2.77×10^{-1}
Hydraulic diameter	(m)	1.08×10^{-1}	1.22×10^{-2}	8.85
Bypass area	(m²)	7.2×10^{-5}	1.54×10^{-2}	4.67×10^{-3}
Hot leg				
Inner diameter D	(m)	5.25×10^{-2}	7.35×10^{-1}	7.13×10^{-2}
Length L	(m)	2.0	7.28	2.75×10^{-1}
$L/\sqrt{D}$		8.72	8.48	1.03
U-tube in one steam generator				
Number		30	5626	5.33×10^{-3}
Avg. length	(m)	4.08	16.85	2.42×10^{-1}
Inner diameter	(mm)	15.4	15.4	1.0
Volume	(m³)	2.28×10^{-2}	18.44	1.23×10^{-3}
Cold leg				
Inner diameter D	(m)	5.25×10^{-2}	7.87×10^{-1}	6.67×10^{-2}
Length L		5.0	15.7	3.18×10^{-1}
$L/\sqrt{D}$		21.8	17.69	1.22
Downcomer				
Flow area	(m²)	0.0185	2.63	7.03×10^{-2}
Hydraulic diameter	(m)	4.12×10^{-2}	4.8×10^{-1}	8.58×10^{-2}
Pressurizer				
Volume	(m³)	9.32×10^{-2}	39.64	2.35×10^{-3}
Surge line flow area	(m²)	3.44×10^{-4}	6.38×10^{-2}	5.39×10^{-3}

3. SCALING CRITERIA AND EXPERIMENTAL CONDITIONS

Following LOFW scenarios, the most important consideration in deriving appropriate scaling criteria is to preserve the power history and the distribution of RCS mass inventory during the transients. Power-to-volume scaling (Nahavandi, 1979) is frequently used to preserve time, power, and mass inventory for full-height full-pressure (FHFP) facilities and for prototype plant because of the same fluid properties at the full pressure. But for RHRP facilities, power-to-volume scaling can not preserve the same relationship of power history and RCS mass inventory as prototype plant because of the different operating pressure. So a power-to-mass scaling was proposed and validated in the study of RHRP IIST and FHFP LSTF/BETHSY counterpart tests on PWR small break LOCA (Lee, 1995).

The experiments were designed to simulate long-term RCS responses including the secondary boil-off and the primary pressurization. So, the core power decay and the pump coastdown during the initial phase of LOFW were not simulated. Based on the power-to-mass scaling, the ratios of power, HPI injected mass flow rate, and initial mass inventory between IIST facility with those of Westinghouse Maanshan PWR were maintained at the same value of 2.5×10^{-3}. The identical initial conditions of two LOFW experiments conducted at IIST are listed in Table 2. The boundary conditions and experimental procedures of the two experiments are described as follows:

The first LOFW experiment was to simulate the RCS behavior with the assumptions of a complete loss of feedwater associated with RCS pump trip, HPI failure, and no accident management by the operator. A core power of 130 kW was set to meet the requirement of steady-state initial conditions. The scaled decay power drops to 130 kW after the reactor scram at 4000 s. Therefore, during the experiment the constant core power was maintained until reaching 4000 s and then the core power had followed the scaled decay power. The core power supply will be tripped by the occurred core uncovery associated with a cladding temperature higher than 873 K.

Using similar experimental conditions as mentioned above, the second LOFW experiment was operated with the application of the bleed-and-feed procedures, open PORV and start HPI flow manually, when any two SG secondary wide range levels were dropped to 6% equivalent to level of 213 mm in the SG at IIST facility. In order to get the simplified boundary conditions for simulating the RCS bleed and feed, three PORVs at the Maanshan PWR were simulated by one PORV at IIST facility with an orifice of 6.1 mm inside diameter to provide an equivalent flow area to that of the three Maanshan PORVs. The characteristics of IIST HPI flowrate is simulated the same as that of the Maanshan PWR by using a flow control system. The experiment was terminated when the validation of bleed-and-feed procedures are verified for the LOFW accident management.

4. EXPERIMENTAL RESULTS AND DISCUSSIONS

The time dependence of 19 groups of measured data in LOFW experiments without operator actions and with bleed and feed are shown in Figs. 2-4 and Figs. 5-7, respectively. Comparison of the transient responses of primary pressure, pressure vessel collapsed liquid level, and the cladding temperature between two LOFW experiments without operator action and with bleed and feed is shown in Fig. 9.

Table 2   The initial conditions of IIST LOFW experiments

Parameter		IIST
Core power	(kW)	130
Primary pressure	(MPa)	1.89
Primary mass inventory	(kg)	444
Hot leg temperature	(K)	471
Cold leg temperature	(K)	434
PZR level/PZR height	(mm)	1806/2930
Secondary pressure	(MPa)	0.58
SG 2nd level/SG 2nd height	(mm)	1540/2950

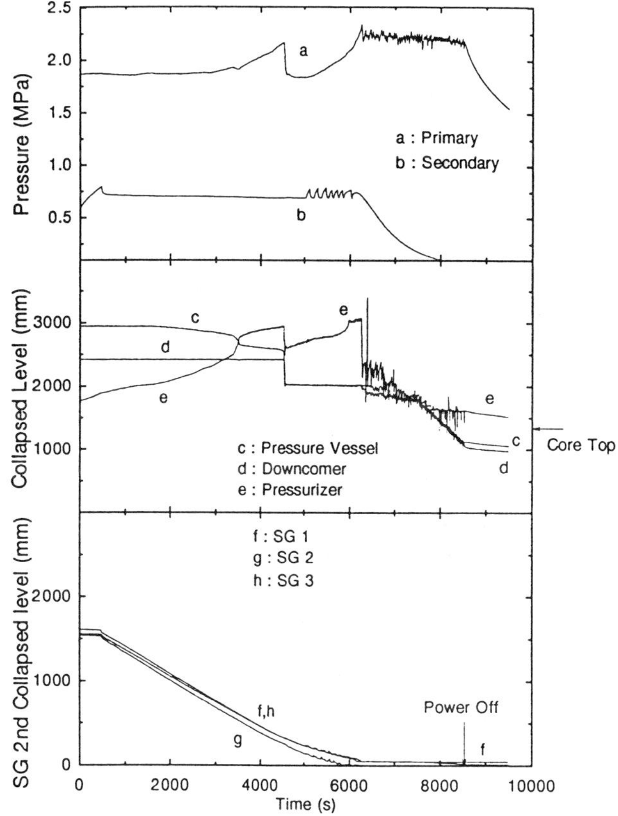

Fig. 2   The time dependence of 8 groups of measured data in LOFW without operator action

4.1 LOFW experiment without operator action

The scenarios and the controlling phenomena of LOFW transient without operator action are exhibited in Figs. 2-4. Based on the sequence in which the different thermal hydraulic phenomena occur, the scenarios are divided into four periods:(1) SG secondary boil-off, (2) primary pressurization, (3) PORV cycled to open and close automatically, (4) core uncovery.

Period I: SG secondary boil-off (from 0 s to 3000 s)

After LOFW, the SG secondary side pressure (Fig. 2- b) increased to the setpoint of the relief valve at 470 s. Then the secondary side levels (Fig. 2-f, g and h) decreased continuously as the liquid inventory boiled away through the relief valve. During this period from 0 s to 3000 s, the natural circulation flow rates of the three loops were about the same (Fig. 4-a, b and c). The primary pressure (Fig. 2-a) remained constant because the core decay heat through the natural circulation was completely removed by the heat sink of SG secondary boil-off.

Period II: Primary pressurization (from 3000 s to 4650 s)

The primary pressure increased due to the degradation of SG heat transfer when the secondary level was dropped below 750 mm after 3000 s. During this period, the simultaneous rise of PZR liquid level (Fig. 2-e) and the drop of the pressure vessel liquid level (Fig. 2-c) resulted from steam voids swelling the coolant from pressure vessel to flow into the PZR. The temperatures of the hot leg reached saturation condition by the occurrence of the swelling phenonena in the pressure vessel are shown in Fig. 3-e and f.

Period III: PORV cycled to open and close automatically (After 4560 s)

The PORV began to open for the first time at 4560 s and a setpoint pressure of 2.21 MPa. A large amount of primary liquid inventory was lost through the open PORV. Therefore, the primary pressure, PZR level and pressure vessel level (Fig. 2-a, e, and c) dropped sharply. After 6160 s, the flow discharged from the cycled open of PORV became a single-phase steam flow and PZR level was obviously lower than the top of PZR. The pressure drops of SG U-tube upflow and downflow side (Fig. 3-a and b), the liquid levels of SG inlet and outlet plena (Fig. 3-c and d) dropped to zero at 7000 s due to the loss of RCS inventory through the discharge of PORV. The pressure vessel level and downcomer level (Fig. 2-c and d) decreased continously but the PZR level was maintained at constant value of 1.5 m (Fig. 2-e) which had indicated that flooding phenomena occurred when the surge line of PZR to holdup liquid in the PZR. The natural circulation was stopped when the mixture level of the pressure vessel became lower than the elevation of hot/cold legs shown in Fig. 4.

Period VI: Core uncovery (After 8300 s)

A loss of secondary heat sink associated with continual loss of coolant through PORV led to core uncovery at 8300 s. In the meantime, the cladding temperature increased sharply to 800 K and the power was tripped (Fig. 3-h) to protect the integrity of heating elements.

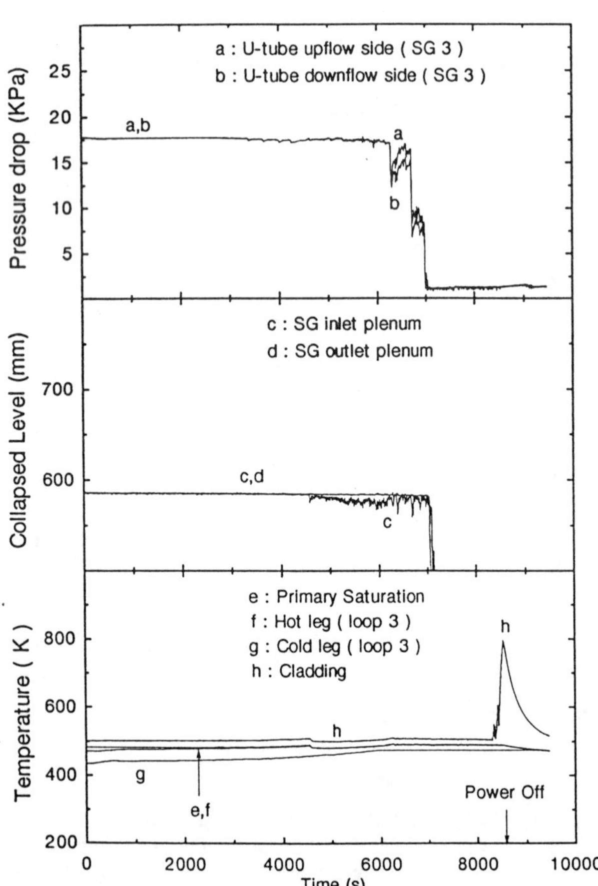

Fig. 3 The time dependence of 8 groups of measured data in LOFW without operator action

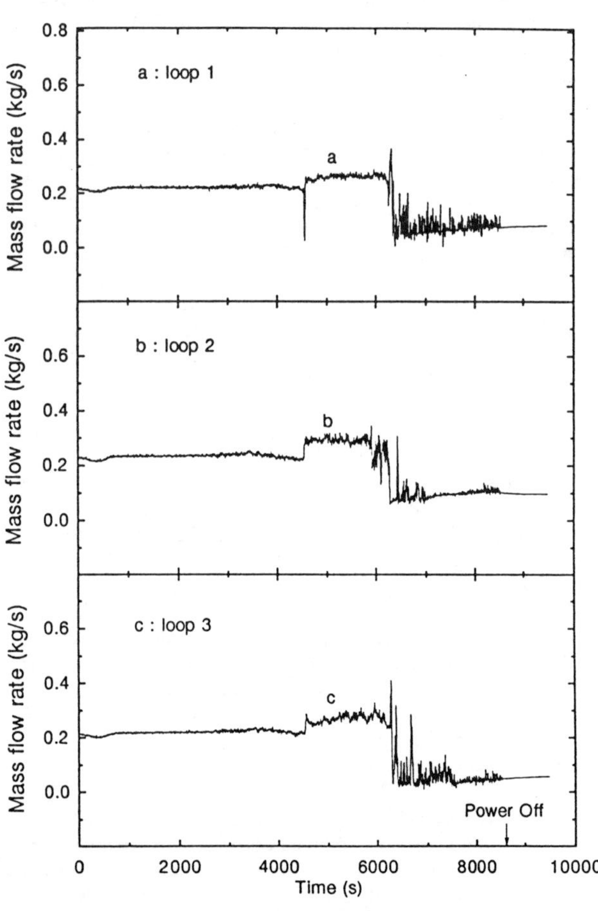

Fig 4 The time dependence of 3 groups of loop mass flow rate in LOFW without operator action

4.2 LOFW experiment with bleed and feed

The scenarios of LOFW with bleed-and-feed transient can be separated into three periods: (1) SG secondary boil-off, (2) primary pressurization, and (3) bleed and feed as shown in Figs. 5-8. Before the application of the bleed-and-feed procedures, the time sequence at the occurrences of SG secondary boil-off and primary pressurization are similar to those of Period I and II in LOFW experiment without operator action. The period III of RCS response with bleed and feed is illustrated as follows. The bleed-and-feed procedures with latched open PORV and initiated HPI were performed when the secondary levels of three SGs (Fig. 5-f, g and h) were dropped to 213 mm at 4500 s. At the initial phase (4500 s to 5000 s) of bleed and feed, the primary pressure, PZR level, pressure vessel level, and downcomer level decreased steeply, as shown in Fig. 5 because the discharge flow through PORV was obviously larger than the feed flow rate from HPI, as shown in Fig. 8. At the same time, the oscillating decrease of cold leg temperature (Fig. 6-g) was due to the mixing of cold HPI fluid with hot RCS inventory in the cold legs.

After 5000 s, the level at pressure vessel, downcomer and PZR became stable (Fig. 5-c, d and e) resulting from the balance of the PORV discharged flow and the HPI feed flow (Fig. 8-a and b). The natural circulation flow rates in each loop were terminated by the actions of bleed and feed as shown in Fig. 7. The removal of core decay heat by natural circulation was replaced by the RCS bleed and feed to maintain long-term core cooling.

4.3 Comparison of experimental results for LOFW with and without bleed and feed

In the comparison of experimental results for LOFW without operator action and for LOFW with bleed and feed as shown in Fig. 9, the RCS responses including SG secondary boil-off, primary pressurization, pressure level decrease, and PZR level increase are very similar during the time period from 0 s to 4500 s. For the LOFW experiment without operator action, the loss of RCS inventory associated with primary pressure equal to or greater than the setpoint of PORV ended with core uncovery at 8524 s. For LOFW experiment with bleed and feed, the pressure vessel level ended as stable and the primary pressure decreased continuously to maintain a long-term core cooling.

The flooding phenomena were observed at the surge line of PZR when the PORV had opened in both LOFW experiments. The core level depression was observed only in LOFW experiment with bleed and feed actions (Fig. 5-c and d). The reason is that the downcomer pressure decreases due to the mixing of cold HPI flow with the steam from the upper plenum via core bypass into the downcomer.

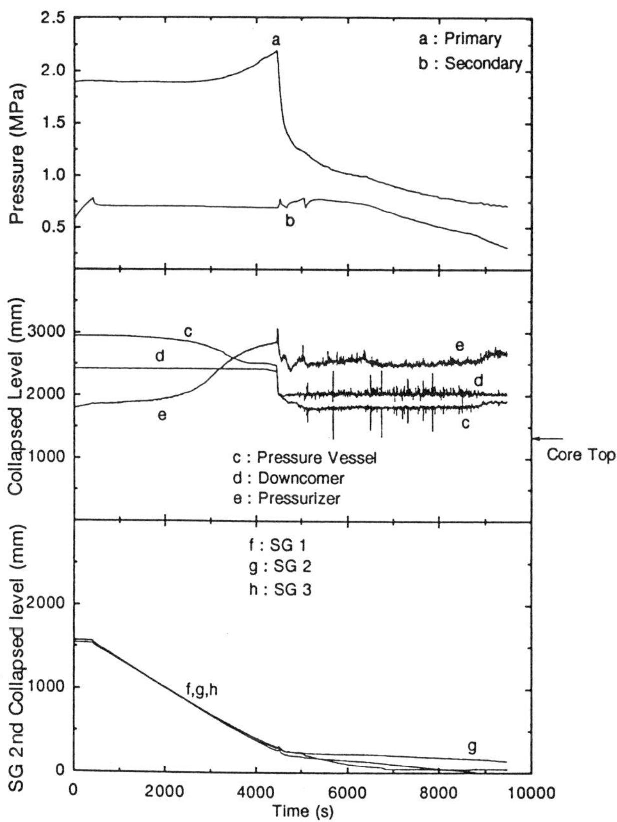

Fig. 5　The time dependence of 8 groups of measured data in LOFW with bleed and feed

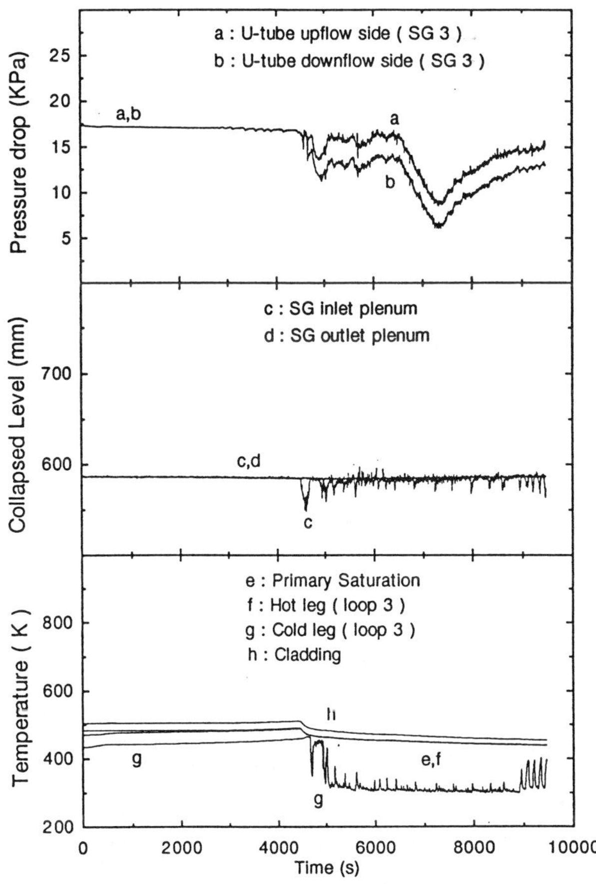

Fig. 6　The time dependence of 8 groups of measured data in LOFW with bleed and feed

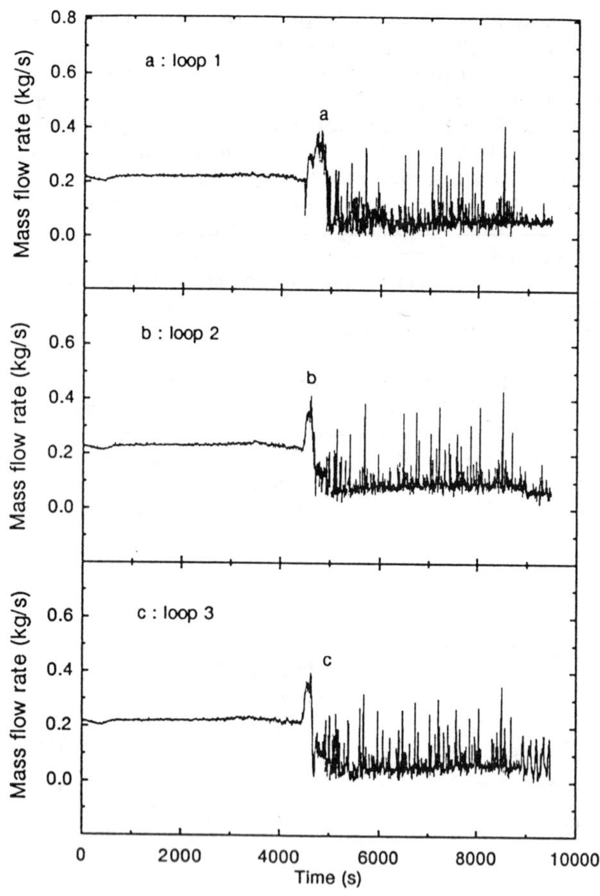

Fig. 7　The time dependence of 3 groups of loop mass flow rate in LOFW with bleed and feed

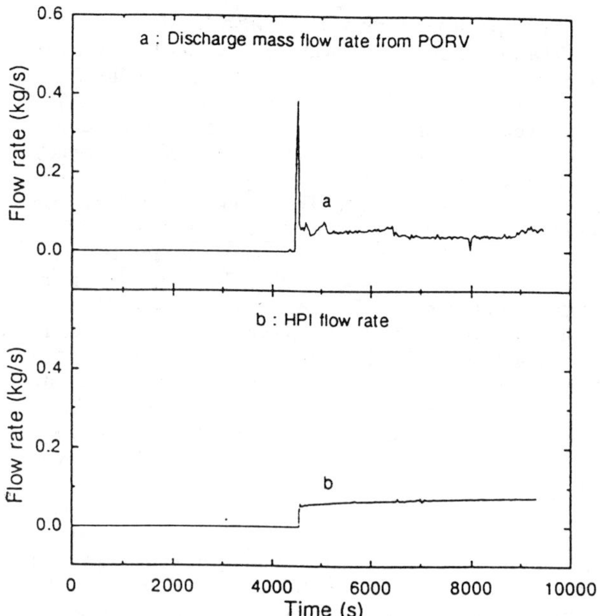

Fig. 8　The time dependence of discharge mass flow rate and HPI flow rate in LOFW with bleed and feed

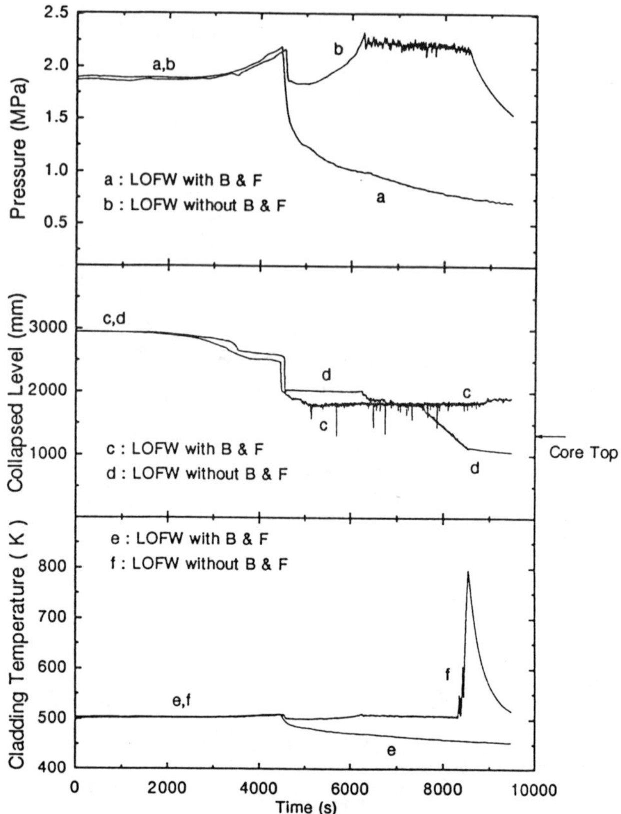

Fig. 9　Comparison of primary pressure, pressure vessel level, and cladding temperature in two LOFW experiments with no operator action and with bleed and feed

5. CONCLUSIONS

Two experiments simulating LOFW without operator action and with bleed and feed actions have been conducted at IIST facility to identify the RCS T/H behaviors and to assess the validity of emergency operating procedures. The major experimental results are summarized as follows:

(1) LOFW without operator action leads to loss of secondary heat sink and loss of RCS mass inventory through PZR PORV. Coreuncovery occurs at 8524 s. Following emergency operating procedures of bleed and feed with HPI operation and PORV open can provide sufficient core cooling for LOFW event.

(2) The liquid holdup in PZR observed from both LOFW experiments is due to the flooding phenomena in the PZR surge line. The core level depression observed in LOFW with bleed and feed is triggered by the condensation of the downcomer steam as the highly subcooled water of HPI flows into the downcomer.

28

REFERENCES

Asaka, H. and Y. Kukita, Primary feed-and-bleed experiments at ROSA-IV LSTF, The Sixth International Topical Meeting on Nuclear Reactor Thermal Hydraulics (NURETH-6), Grenoble, France October 5-8, 1993.

Billa, C., et al., Accident management for a loss of feedwater in a PWR, Nuclear Technology, vol. 98, June 1992.

Chataing, T., et al., Total loss of feedwater on the BETHSY integral test facility, Proceedings of Safety of Thermal Reactors, International Topical Meeting, Portland, Oregon, July 21-25, 1991.

Lee, C.H., et al., The conceptual design report for INER integral system test facility, INER-1098, 1991.

Lee, C.H., et al., IIST and LSTF/BETHSY counterpart tests on PWR small break LOCA, International Symposium on Two-phase Flow Modeling and Experiment, Rome, Italy, October 9-11, 1995.

Nahavandi, A.N., et al., Scaling laws for modeling nuclear reactor system, Nuclear Science and Engineering, 72, pp. 75-83, 1979.

Umminger, K. et al., Station blackout experiment in the PKL III test facility and RELAP5/MOD2 analyses, International Conference on Nuclear Engineering (ICONE-3), Kyoto, Japan, April 23-27, 1995.

Loss of Residual Heat Removal System Simulation
Using CATHARE Thermal Hydraulic Code

A. A. Troshko

Y. A. Hassan

Department of Nuclear Engineering

Texas A&M University

College Station, Texas 77843-3133

Abstract

Thermal hydraulic CATHARE V1.3U code has been used to simulate an International Standard Problem (ISP38) experiment conducted at BETHSY Integral Test Facility located in Grenoble, France.

This experiment dealt with simulation of the loss of Residual Heat Removal System (RHRS) during midloop operation. It involved opening of the pressurizer manway and steam generator outlet plenum manway simultaneously with switching on the heating rods power simulating loss of RHRS. The total power was kept unchanged with the level 138 kW throughout the test. Mass discharge through both manways led to core boiling and uncovery. Test was stopped when the primary cooling system was filled back to a midloop level.

Overall, the code's prediction and experimental data were found to be in a reasonable qualitative agreement. However, the code underestimated the time of the core uncovery and the actuation of the gravity feed injection due to the ovepredicted mass discharge through steam generator manway during the initial stage of the transient. This was caused by CATHARE's miscalculation of the phase separation effect at the hot leg/surge line tee junction and significant water entrainment into the surge line at the beginning of the test. It was found that the code's model of the upward tee junction needs to be refined for the low pressure ranges.

1. Introduction

Most Pressurized Water Reactor (PWR) safety studies have traditionally addressed the problem of accident transients associated with power operation of the plant. However, recent experience has included several events during outages after plant shutdown, and probability risk assessments show that the risk of core damage due to the events occurring during shutdown cannot be ignored as it is estimated to be of the same order of magnitude as the risk from accidents during full power operation.

After shutdown, reactor cooling is achieved by the Residual Heat Removal System (RHRS) and loss of RHRS's capability is, therefore, considered to be the primary cause for possible core damage during the cooling mode. The greatest risk is associated with so-called midloop operation which is mainly characterized by a primary cooling system drained to the hot leg (HL) centerline and the existence of openings (vent paths and/or manways) through which a further reduction of the mass inventory is to be expected due to boil and spill over of liquid. According to the size and the location of the primary openings, core uncovery may be reached more or less rapidly, and core damage may occur if adequate countermeasures are not undertaken in time (Dumont et al., 1993).

The loss of the RHRS during shutdown involves conditions which differ significantly from those which prevail in accident situations initiated at full power operation of the reactor. Typically, core power is very low (less than roughly 0.6% of nominal power), primary pressure is usually close to atmospheric pressure (especially when one or several manways are open) and large amounts of air or nitrogen are present in the upper elevations of the primary loops. Capabilities of existing thermal hydraulic codes in these particular conditions are questionable, and it is necessary to provide data against which the performance of codes may be assessed.

The objective of the present study is to perform a simulation of the 6.9c (ISP38) loss of RHRS experiment conducted at BETHSY facility using the French thermal hydraulic code CATHARE. The primary aim of the simulation is to assess and validate the computer code's ability to predict accurately important physical phenomena during the accident scenario of the loss of the RHRS.

2. BETHSY Integral Test Facility

BETHSY is an integral test facility located at the Nuclear Center of Grenoble, France, whose purpose is the investigation of PWR accident transients. Its design aims at answering the need for an adequate representation of the commercial PWR in order to be capable of simulating most accident situations of interest while minimizing the distortions of relevant physical phenomena.

The selected reference Nuclear Power Plant is a three loop, 900 MWe (2775 MWt) Framatome PWR, and the BETHSY facility includes all the corresponding circuits and systems which are likely to play a role in case of transient as far as thermodynamic aspects are concerned. BETHSY is a full pressure facility. The maximum operating pressure of the primary coolant system is 17.2 MPa while the secondary side can withstand pressures up to 8 MPa. The scale selected for the volumes of the facility is f=1/100 (more precisely $1.032*10^{-2}$).

A scaling factor of 1/1 for elevations of every component has been chosen to simulate such phenomena as natural circulation and gravity driven flows.

3. CATHARE code description

CATHARE (Code for Analysis of Thermalhydraulics during an Accident of Reactor and Safety Evaluation) has been developed to perform best estimate calculations of pressurized water reactor accidents: PWR loss of coolant (large or small break, primary and secondary circuits).

CATHARE has been developed by the French Atomic Energy Commission (CEA), Electricite de France (EDF), and Framatome and includes several independent modules that take into account any two phase flow behavior:

1)   mechanical non-equilibrium vertical: co- or countercurrent flow, flooding countercurrent flow limitation (CCFL), etc. horizontal: stratified flow, critical or not critical flow, co- or countercurrent flow, etc.

2)   thermal non equilibrium. critical flow, cold water injection, super-heated steam, reflooding, etc.

3)   heat transfer from wall to fluid and from wall to vapor, etc.

In order to take into account these phenomena, the CATHARE code is based on a one-dimensional, two-fluid, six-equation model with a unique set of constitutive laws. The code is limited to transients during which no severe damage occurs to fuel rods. Fuel ballooning and clad rupture are supposed to have no major effect on water flow in the primary circuit.

In order to obtain the minimum computing time, effort was given to achieving the longest possible time step size which led to a fully implicit numerical scheme (Farvacque and Sarrette, 1992).

4. Description of the loss of RHRS test

The aim of this particular test was to study the accident transient following a failure of the RHRS with the pressurizer and steam generator 1 (SG1) outlet plenum manways open. The facility was in its normal conditions. In particular, the pressurizer was connected to loop 1. Neither the normal nor auxiliary spray systems were used. With the exception of the thermal barriers, the pump cooling circuits were not used. The SG secondary sides were full of air and isolated.

On the upper head-downcomer link, the valve was fitted to the tubular part of the upper head bypass. The valve was calibrated in its standard position. Its aim was to reduce the flow rate simulating real PWR conditions. The orifice representing the pressurizer manway was located 669 mm above the top of the pressurizer. The manway axis was vertical and its diameter was 41 mm.

The orifice representing the SG1 outlet plenum manway was located on the vertical part of the intermediate leg1 (IL1). Its axis, which was horizontal, was located 720 mm above the axis of the hot legs. Its diameter is also 41 mm.

The trace heating system has been employed in the experiment in order to make environmental heat losses negligible as it is in the case of the real PWR (Lavialle et al., 1995). The trace heating power was maintained constant throughout the test.

Initial conditions were determined during a period of 200 sec prior to the start of the transient. Initially, the liquid level in the primary coolant system (PCS) was close to the hot leg level nozzle (midloop operation). In the upper plenum, the fluid was at atmospheric pressure and with temperature close to saturation. There was no non-condensible gas in the primary system so only steam above the liquid level existed during this test. The steam generators were also unavailable.

Since there is no RHRS in the BETHSY facility, its loss was simulated by increasing the core power which was zero during the steady state. During the test the following actions were performed:

1) at the start of the test the core power was turned on linearly reaching 138kW in 15 seconds (or 0.5% of the nominal power) and the manway orifice valves were opened within 1 sec

2) when the core cladding temperature at the top of the heaters reaches 250 C a gravity feed injection was actuated in the cold leg3

3) the transient was terminated when the mixture level in the pressure vessel reached the hot leg nozzle (this happend at 9044 sec when the hot leg void fraction reached 80%)

The SG secondary systems were full of air and isolated. The lower sections of the SG shell were trace heated. The trace heating was regulated to maintain a wall temperature of 105°C during the steady state and throughout the test. The temperature differences

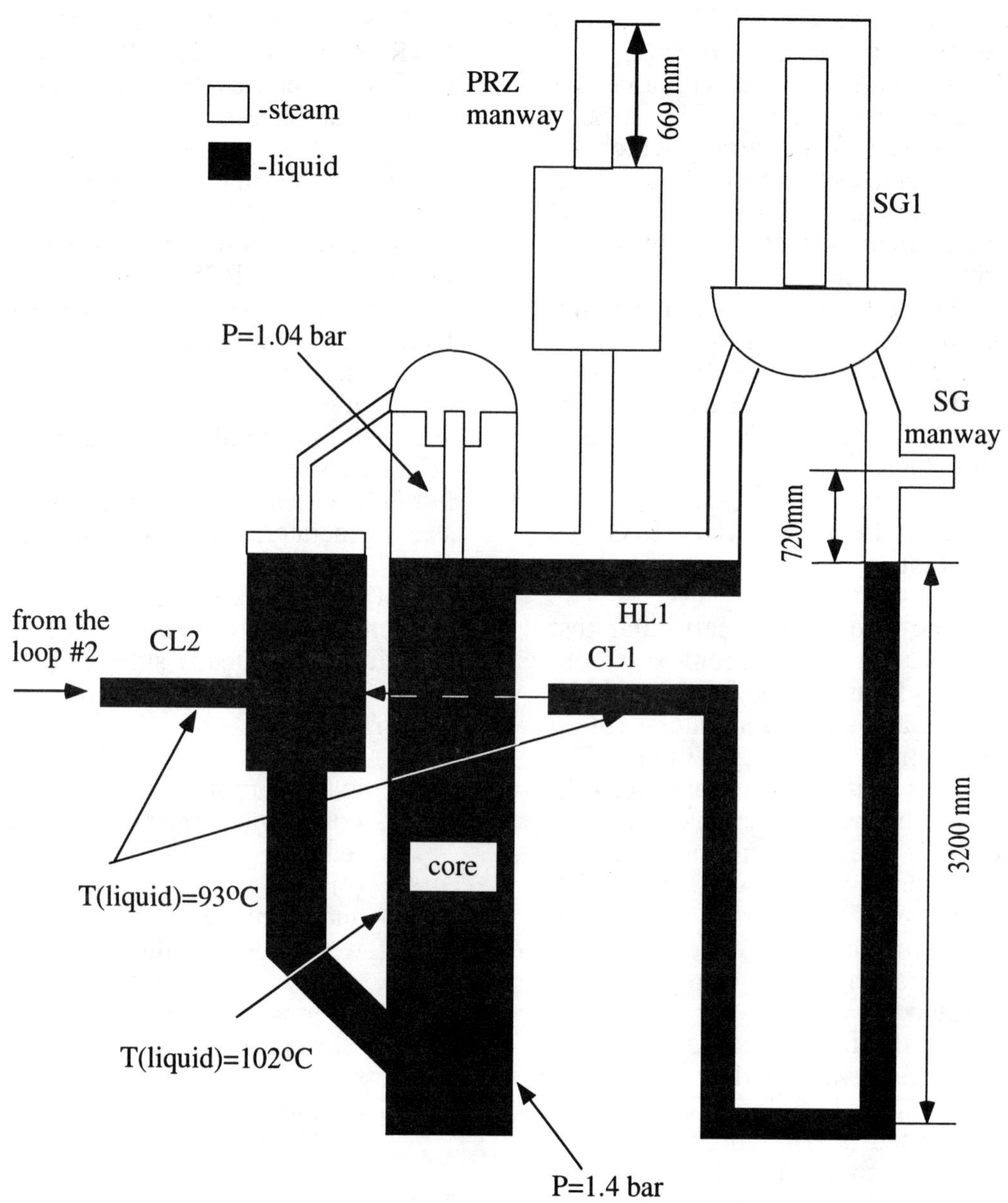

Fig. 1. BETHSY ISP38: transient: prior to the transient.

observed between the primary and secondary side temperatures were too small to be meaningful in relation to the experimental uncertainties. Thus, a satisfactory description of the thermal environment for the steam generator could not be given.

Figure 1 shows the schematic of of the experimental facility configuration prior to the transient initiation. The overall behavior of the transient can be divided into three phases.

1) <u>first phase</u> (from 0 to 2670 sec)
During this phase the two phase level in the pressure vessel is close to the axis of the hot legs. Figure 2 shows the transient conditions at time 1000 sec.

2) <u>second phase</u> (from 2670 to 5660 sec)
Figure 3 shows conditions for the second stage when the two phase level in the pressure vessel falls below the hot leg nozzle level.

3) <u>third phase</u> (from 5660 sec to the end of the test)
During this phase PCS fills to mid-loop level. Figure 4 shows this stage of the transient.

5. Results and Discussions

5.1 CATHARE nodalization scheme of ISP38 experiment

A standard input deck simulating BETHSY facility was obtained from the Centre d'Energie Atomique (France). Although this input deck served as a starting point for the present calculation, several modifications had to be incorporated to make it suitable for the present analysis.

The CATHARE nodalization scheme of the BETHSY facility is shown in Fig. 5. It has 18 pipe type modules, 7 volume type modules, 2 tee type modules and 35 junctions.

5.2 Steady state calculations

In order to model the loss of RHR phenomena accurately, it was necessary to calculate the steady state condition. This was achieved using the experimental boundary and initial conditions.

The primary cooling system (PCS) was filled with the water at a tmperature of 103 C and pressure of 3 bar. The whole system was made isothermal during steady state initialization.

To obtain steady state conditions, the PCS sink was initiated in the lower plenum module until the mass of the system became equal to the experimental mass.

In the next stage of the steady state initialization, the sink was closed and short transient without any trips was run to avoid possible discontinuities in the steady state.

Tables 1 to 3 show comparison of the calculated and experimental (Lavialle et al., 1995) steady state conditions. As shown, calculated characteristics closely coincide with experimental results within experimental error bars.

5.3 Transient results

Once a steady state condition was achieved, a transient was initiated according to experimental boundary conditions.

Figure 6 plots the total mass of the PCS. At time between 500 and 1000 sec, calculated mass continues to decrease while experimental mass steadies due to its redistribution in the system. Such discrepancy between calculated and experimental mass (approximately 180 kg at time 1000 sec) led to earlier calculated core uncovery. Consequently, earlier gravity feed injection was calculated.

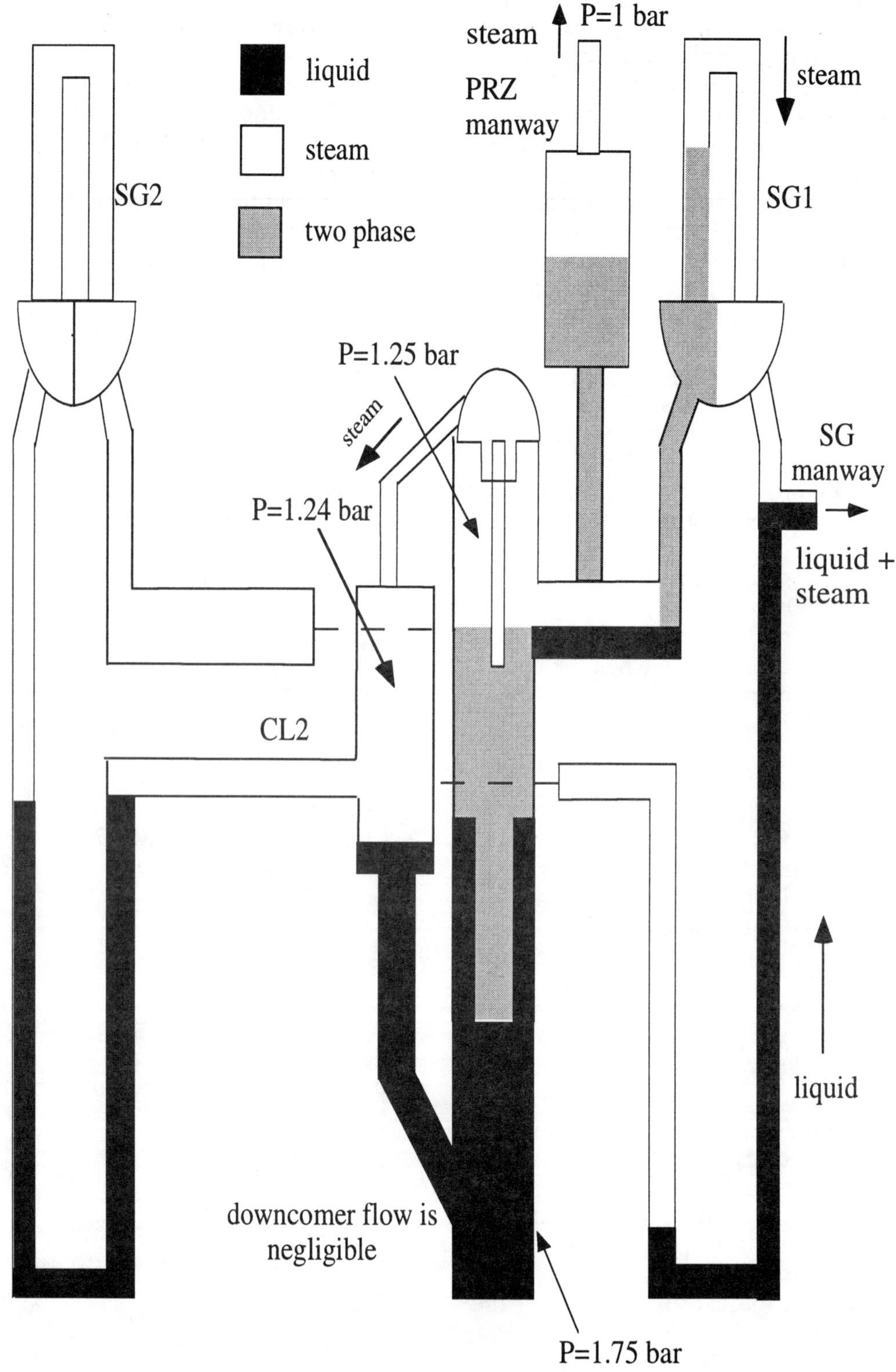

Fig. 2. ISP38 experiment: time = 1000 sec.

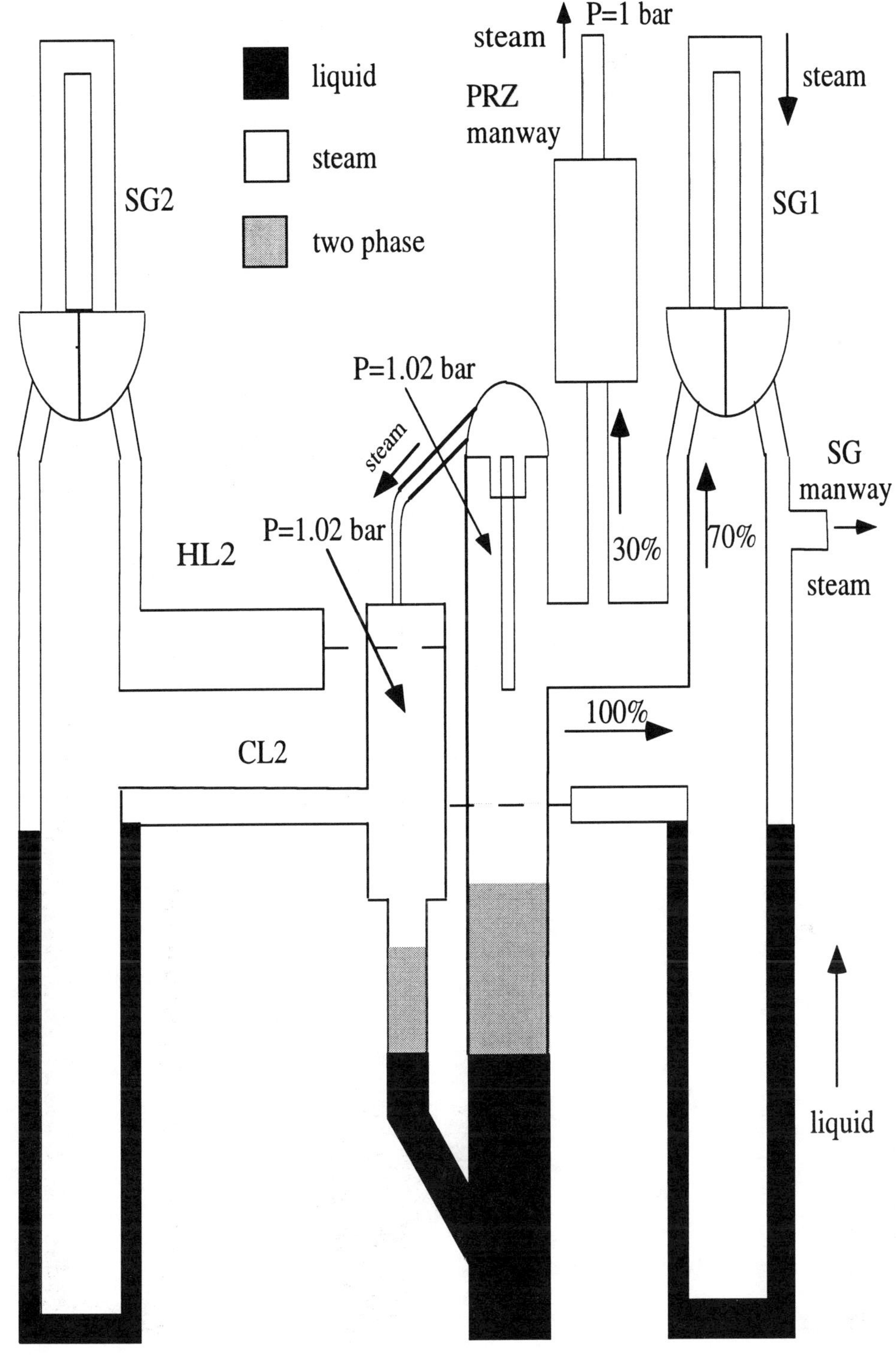

Fig. 3. ISP38 experiment: time = 4000 sec.

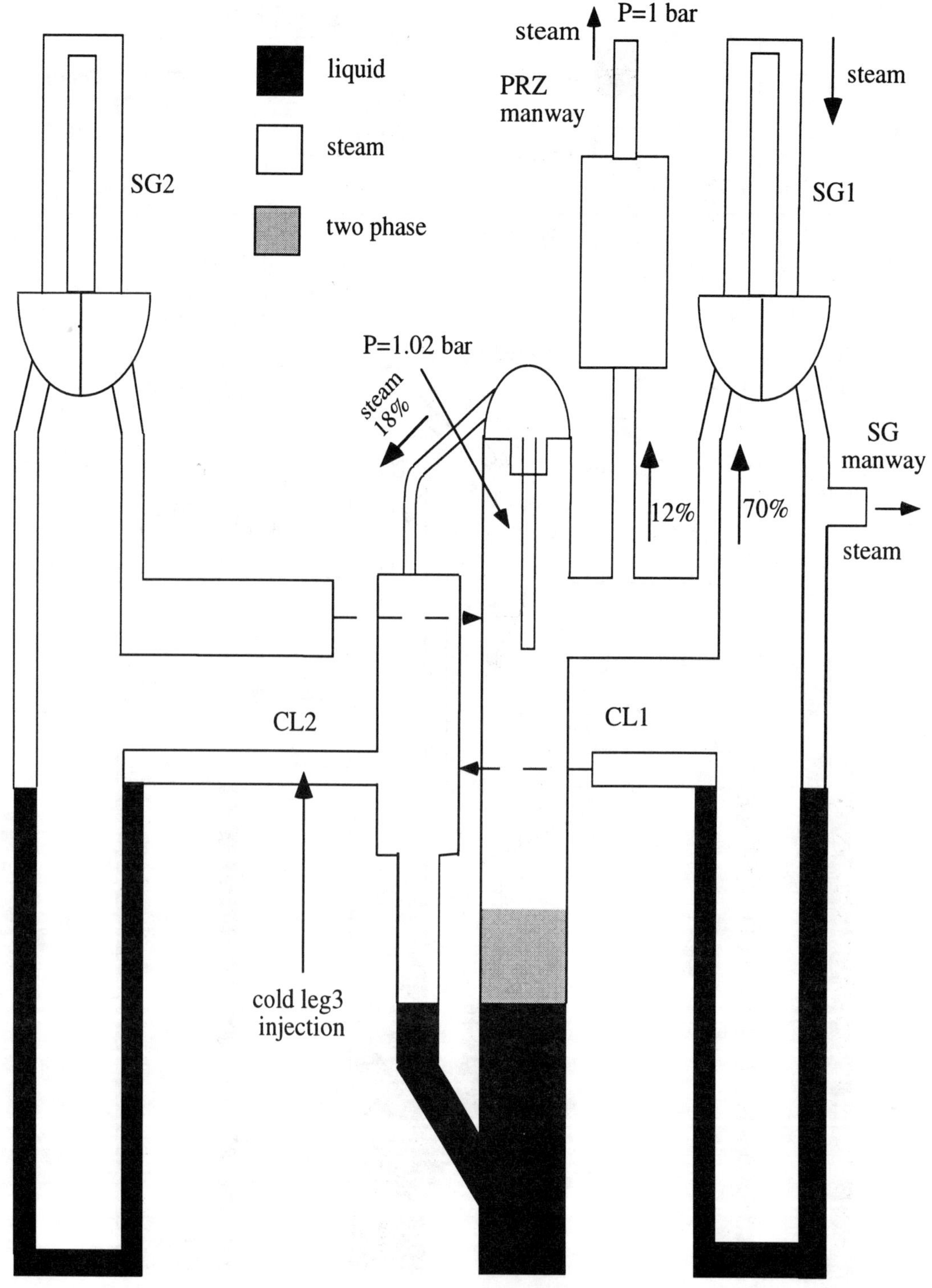

Fig. 4. ISP38 experiment: time = 6000 sec.

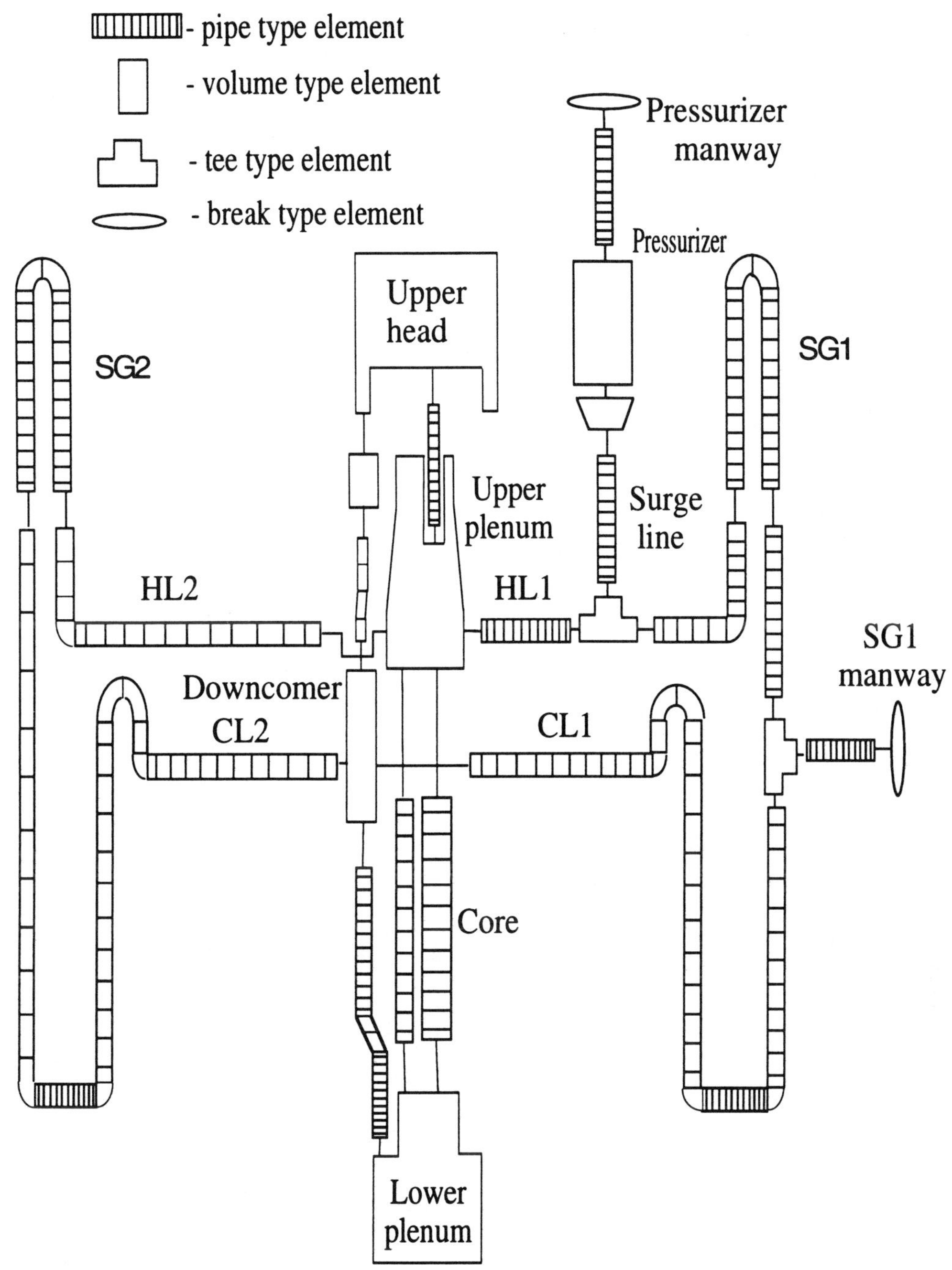

Fig. 5. CATHARE's nodalization scheme for the ISP38 experiment.

Table 4 presents the timing of the main transient events. At the beginning of the transient (from 0 to 1000 sec) as the upper plenum is initially saturated and as soon as the core power is increased and the manways opened, a mixture level rapidly appears in the pressurizer due to the liquid entrainment in the surge line and the vertical parts of

Table 1

PCS initial conditions (pressure vessel)

Data	Experiment	CATHARE
Upper plenum pressure, Pa	$(1.05 \pm 0.02)*10^5$	$1.046*10^5$
Core power, kW	0	0
Primary mass inventory, kg	1068 ± 15	1072
Vessel mass inventory, kg	700 ± 9	696

Table 2

PCS initial conditions (pressure vessel)

	Experiment	CATHARE	Experiment	CATHARE	Experiment	CATHARE
Data	Loop1	Loop1	Loop2	Loop2	Loop3	Loop3
HL void fraction	$0.58 \pm .05$	0.584	$0.55 \pm .05$	0.563	$0.52 \pm .05$	0.563
CL void fraction	$0.00 \pm .05$	0.024	$0.00 \pm .05$	$1.0*10^{-5}$	$0.00 \pm .05$	$1.0*10^{-5}$
Loop mass , kg	124 ± 2	124.65	126 ± 2	126.65	127 ± 2	126.62

Table 3

Wall and fluid average temperatures (Celsius) in PCS (pressure vessel)

	Experiment	CATHARE	Experiment	CATHARE
Data	Fluid	Fluid	Wall	Wall
Upper plenum	102±2	101.6	109±4	100.8
Upper head	104±2	101.6	103±4	101.5
Pressurizer	101±2	100.8	104±4	102.7
Lower plenum	101±2	102.5	104±4	102.5
Core heated length	102±2	102.4	-	-
Downcomer	99±2	102.0	101±4	100.7

the hot legs and the up-flow side of the SG1 U-tubes.

The primary system pressure increase resulted in the following events:

1) an inversion of the pressure difference between hot and cold legs which becomes positive;

2) a drop in the liquid level in the down flow part of the IL 2 and 3 caused by increase in the upper plenum pressure (as shown in Fig. 7) which is the consequence of the pressurizer and surge line behavior;

3) a rise of the water level in the down flow part of the IL1 up to the lower bound of the SG manway discharge line. A two-phase flow took place through the SG manway with steam from the SG1 U-tubes and liquid spilling over from intermediate legs.

During initial phase of the transient, the code qualitatively predicted all main phenomena. However, water mass discharged through the SG manway was overpredicted. Such overprediction was caused by overestimation of the upper plenum pressure peak at the beginning of the test ($P_{exp}^{max} = 1.25\ bar$, $P_{calc}^{max} = 1.31\ bar$). Therefore, the liquid flow from SG1 U-tubes spilling toward SG manway had been predicted. However, the experimental data showed only single phase steam flow throughout the transient.

The first 1000 seconds of the transient (when two-phase flow took place from the SG manway) the liquid discharge from the SG manway was overestimated because of the overeprediction of the upper plenum pressure which governed flow from both manways this is shown in Fig. 8.

Figure 9 presents core swollen level. As it was mentioned, calculated swollen level began to decrease earlier than in experiment which led to earlier calculated initiation of the cladding temperature rise of heaters. This is illustrated in Fig. 10.

Table 4

Main events of the transient

Main events	Experimental time, sec*	Calculated time, sec
Core power turned on	0	0
Core power = 138 kW	15	15
Upper plenum pressure peak	946	628
Two phase flow through the SG1 manway	0-300 and 500-1000	0-794
Cold leg1 is empty	1149	755
Hot leg1 is empty	2672	2962
Upper plenum is empty	3433	2614
Increase in cladding temperature	4620	2790
Minimum core mixture level	5640	3232
Gravity feed injection actuation	5660	3196
Maximum core cladding temperature	6142	3292
Core is reflooded	6805	3720
Hot leg void fraction is 0.8	8868	7474
Experiment is stopped	9044	9044

* - (Lavialle et al., 1995)manway.

The level in the core rapidly increased since the injection flow rate was sufficient to completely reflood core. Test was stopped when the hot leg void fraction was about 80%.

Figure 11 shows mass error of the calculation. The relative mass error of this transient was well below 1%.

5.4 Effect of the surge line/HL1 tee junction

The comparison of the calculated and experimental results reveals that the timing of the main events strongly depended on the system behavior during initial period (from 0 sec to 1000 sec). During this period two-phase flow from SG manway occured. Such flow existed, as long as, pressure in the upper plenum was increasing. Therefore, mass discharge during this period strongly depended on the duration of such increase and its magnitude. The upper plenum pressure behavior, in turn, was influenced by the processes in the surge line and pressurizer. In particular, it strongly depended on the flow pattern in the hot leg1 and surge line and mass discharge characteristics from the pressurizer manway.

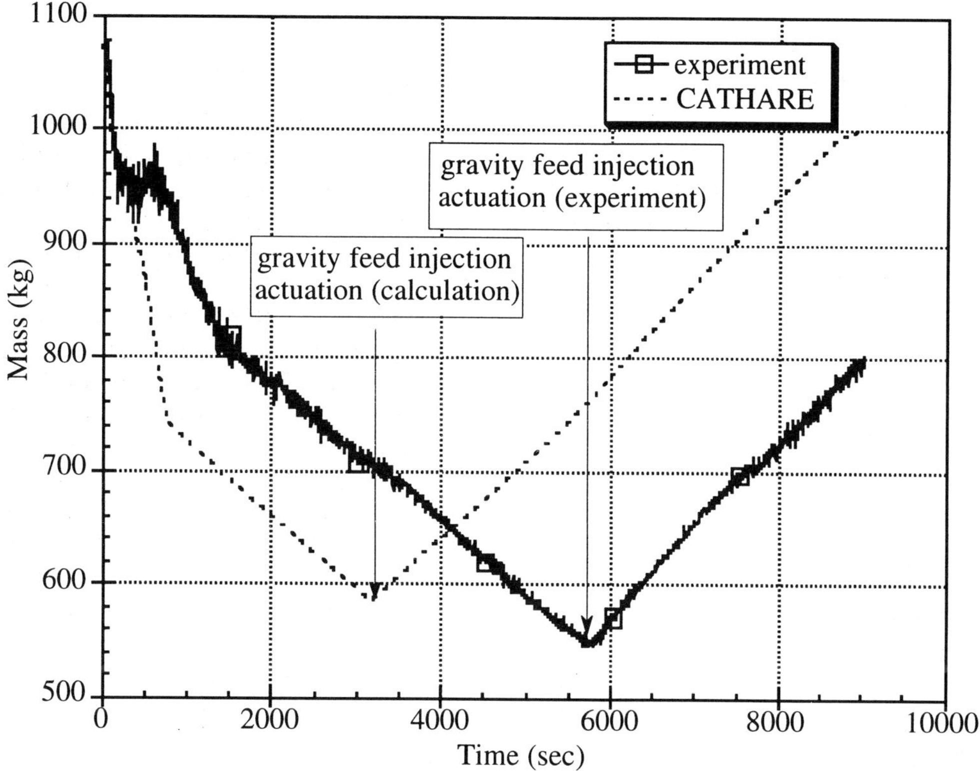

Fig. 6. Total mass of the system

Figure 12 shows the value of the void fraction in the surge line at 0.5 m above its HL1 nozzle during first 1000 seconds of the transient. There is a significant water entrainment for the first 100 seconds, while experimental data show that there is steam single phase flow. Such calculated flow regime in the surge line led to zero mass discharge from pressurizer manway during this period, and this, in turn, resulted in the upper plenum pressure peak in the very beginning of the transient, which was not measured experimentally. Consequently, this pressure peak immediately led to excessive liquid discharge from the SG manway.

To understand possible reasons of the surge line void fraction calculation, the model describing dependence of the tee upward junction on the main pipe void fraction is described below.

According to the present model employed in the present version of the code, when two-phase flow in the horizontal main pipe is non-stratified, the void fraction of the upward tee junction is set to be equal void fraction at the central scalar point of tee module given that liquid velocity in the main pipe is low enough. If it is stratified, then upward junction mass quality will depend on the phase separation level in the main pipe (Janicot, 1992).

In CATHARE's stratification criterion, it is assumed that the phase mixing effects due to liquid turbulence can become so high that they prevent stratification from being set up. This could occur when the turbulent velocity scale V_t is higher than the bubble rise velocity V_b. The approximate value of V_t can be related to the frictional velocity as:

$$V_t \cong \sqrt{\frac{\tau_{wl}}{\rho_l}} = V_l \sqrt{\frac{C_w}{2}} \qquad (1)$$

where ρ_l is a liquid density, V_l is a liquid velocity, τ_{wl} wall shear stress and C_w is a wall drag coefficient.

A simple correlation is used to express V_b :

$$V_b = 1.4 \left[\frac{g \sigma_l (\rho_l - \rho_g)}{\rho_l^2} \right]^{1/4} \qquad (2)$$

where g is gravity acceleration, σ_l is a liquid surface tension and ρ_g is a gas density.

A limitation criterion to stratified flow occurrence is proposed as (Janicot, 1992):

$$V_l < V_b \qquad (3)$$

Figure 13 plots the tee upward junction (surge line/HL1) void fraction as a function of tee central node void fraction during first 500 seconds. Stratification index in the HL1 calculated at 0.5 m from HL1/pressure vessel junction is also shown. From this figure, it can be seen that during first 100 seconds of the transient the flow in the HL1 is non-stratified and void fraction in the upward tee is equal to the tee central node void fraction. This equality originating from CATHARE's donor principle has been validated for high main pipe pressures (from several bars and higher), but it may not be valid for low pressure cases as it was demonstrated here. A refinement in the stratification model is needed.

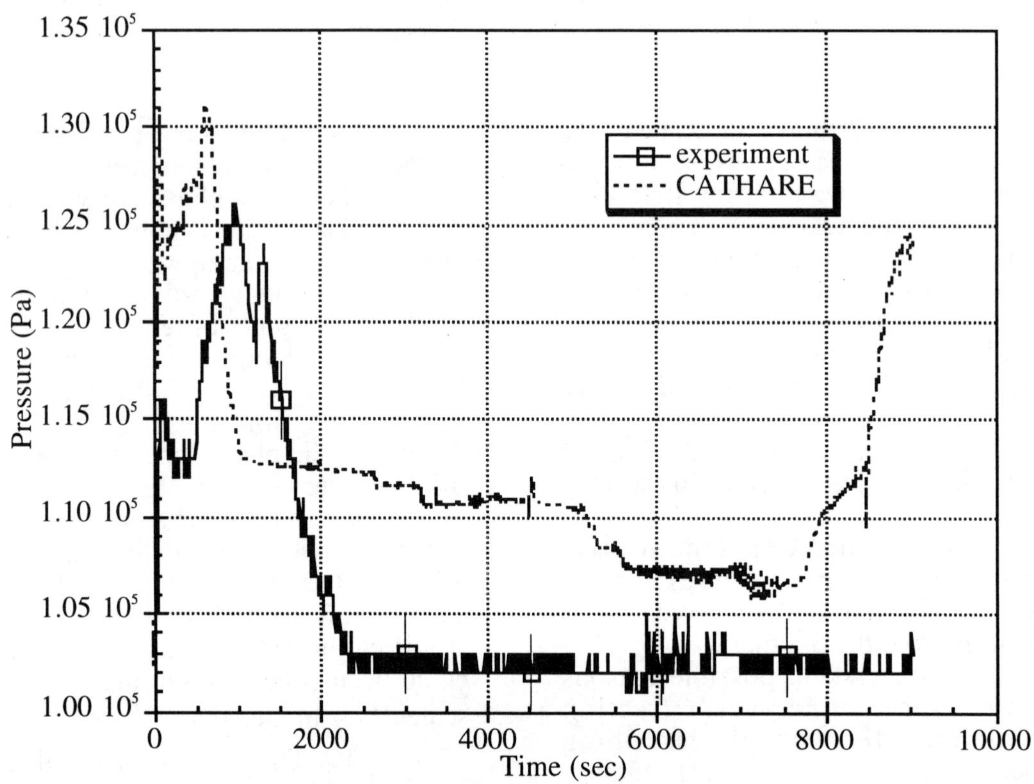

Fig. 7. Upper plenum pressure.

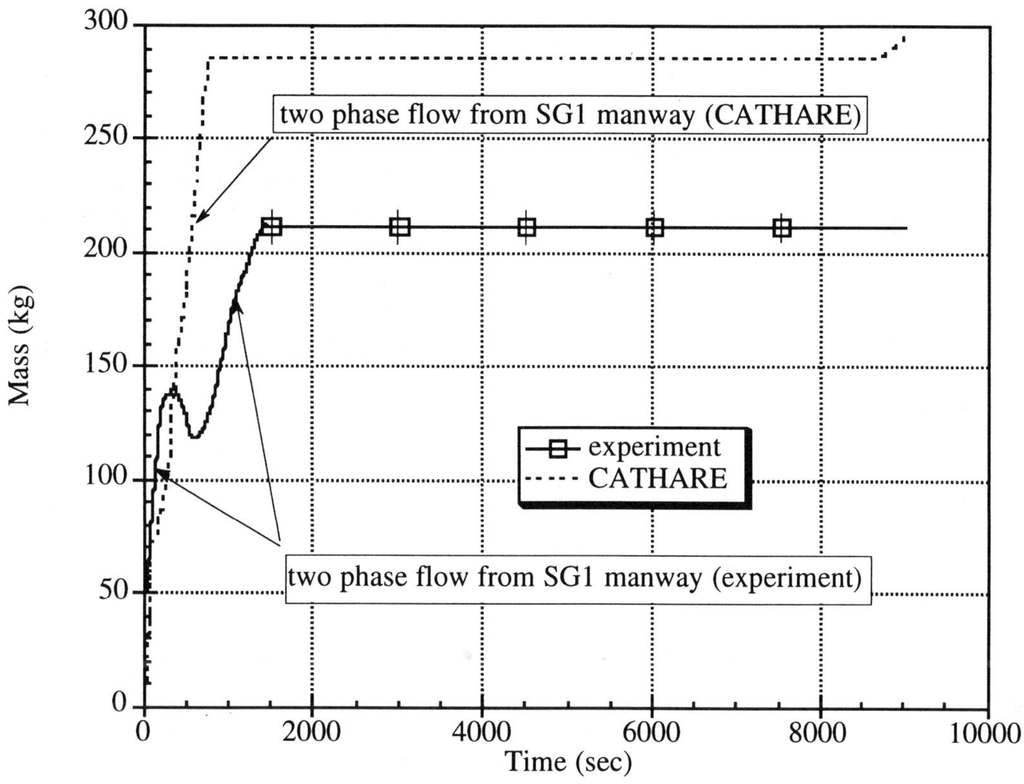

Fig. 8. Time integrated mass flow rate from SG manway (liquid).

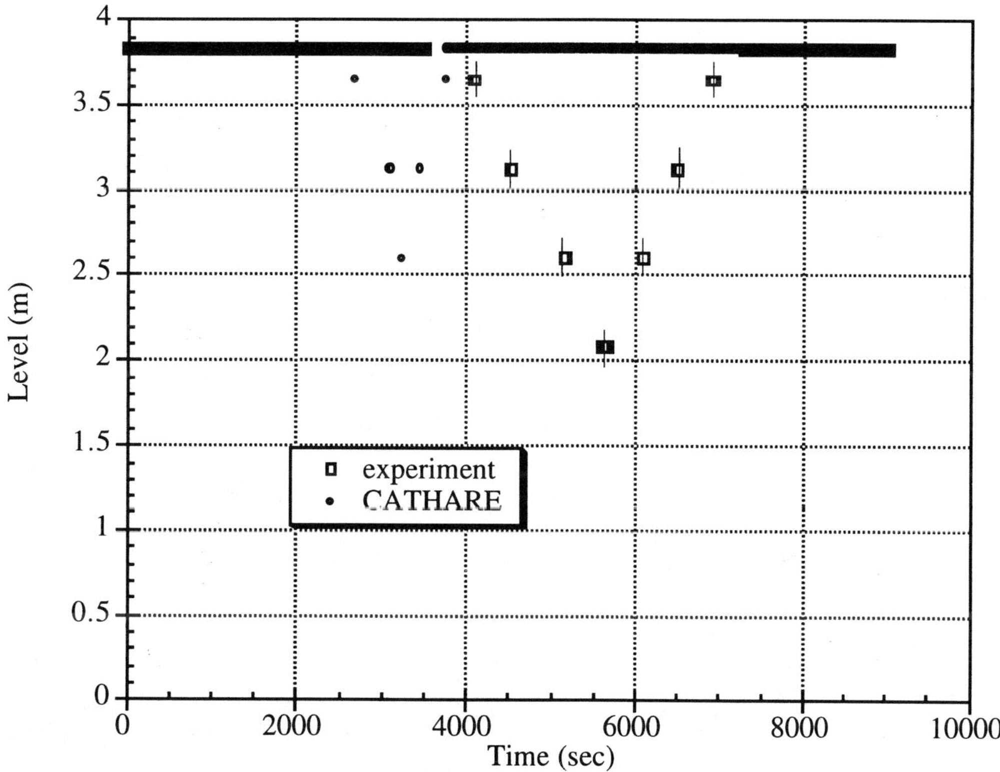

Fig. 9. Core swollen level.

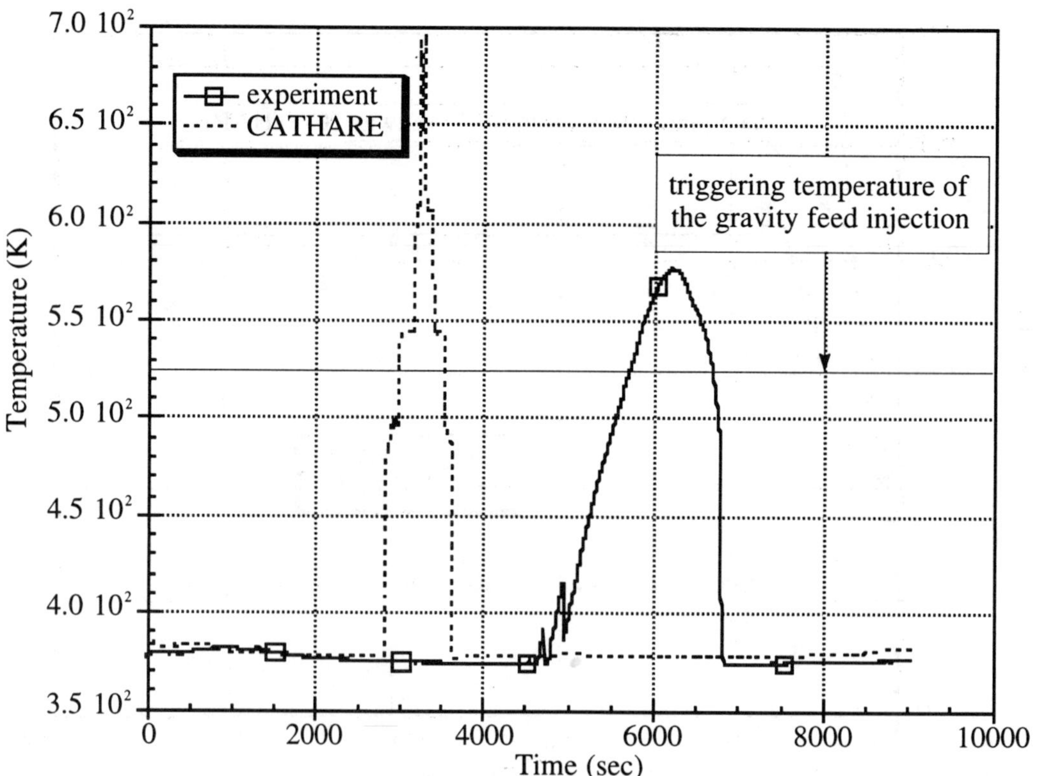

Fig. 10. Heating rod's cladding temperature (top of the heated length).

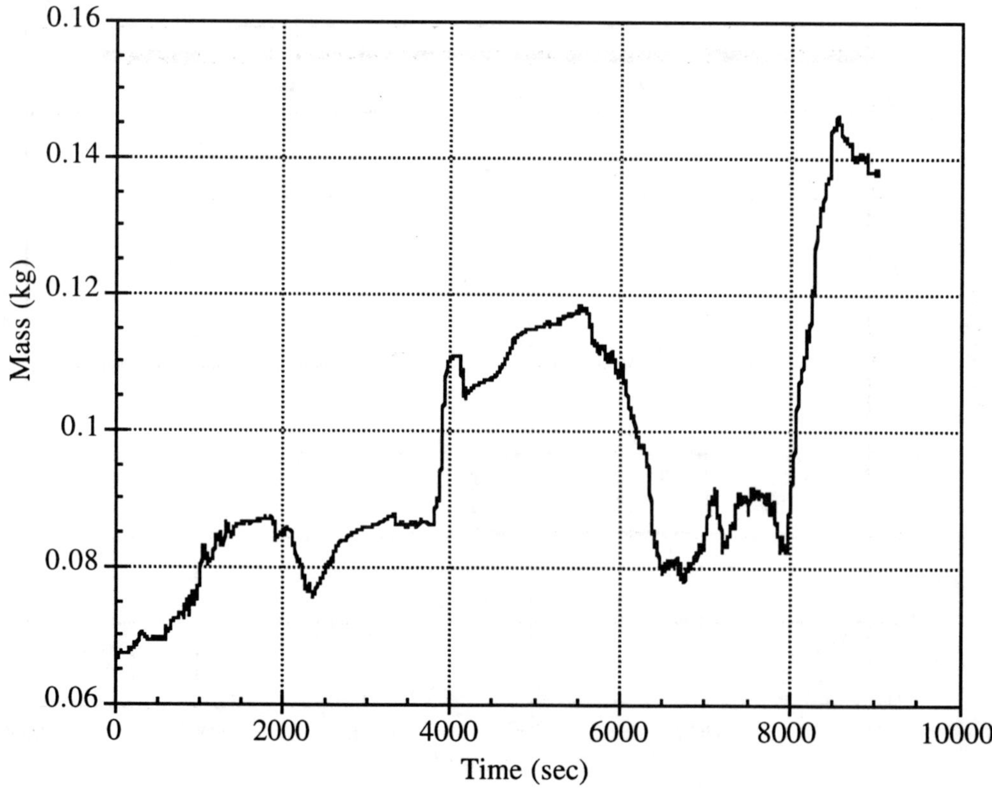

Fig. 11. Mass error in CATHARE calculation.

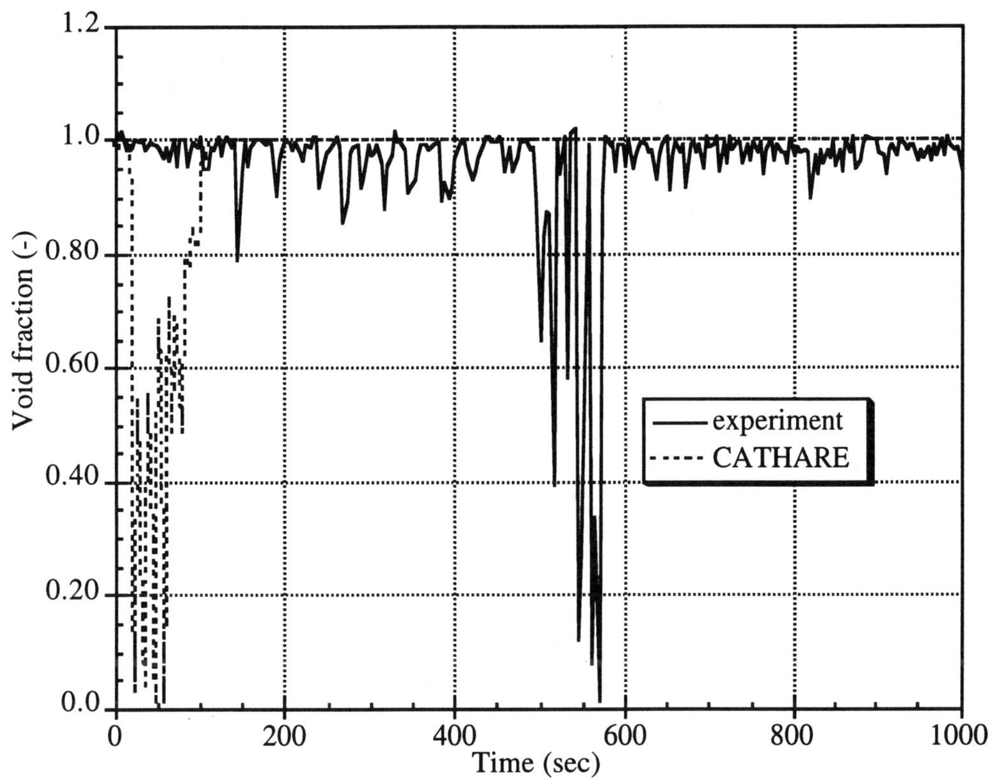

Fig. 12. Surge line void fraction at 0.5 m above its nozzle.

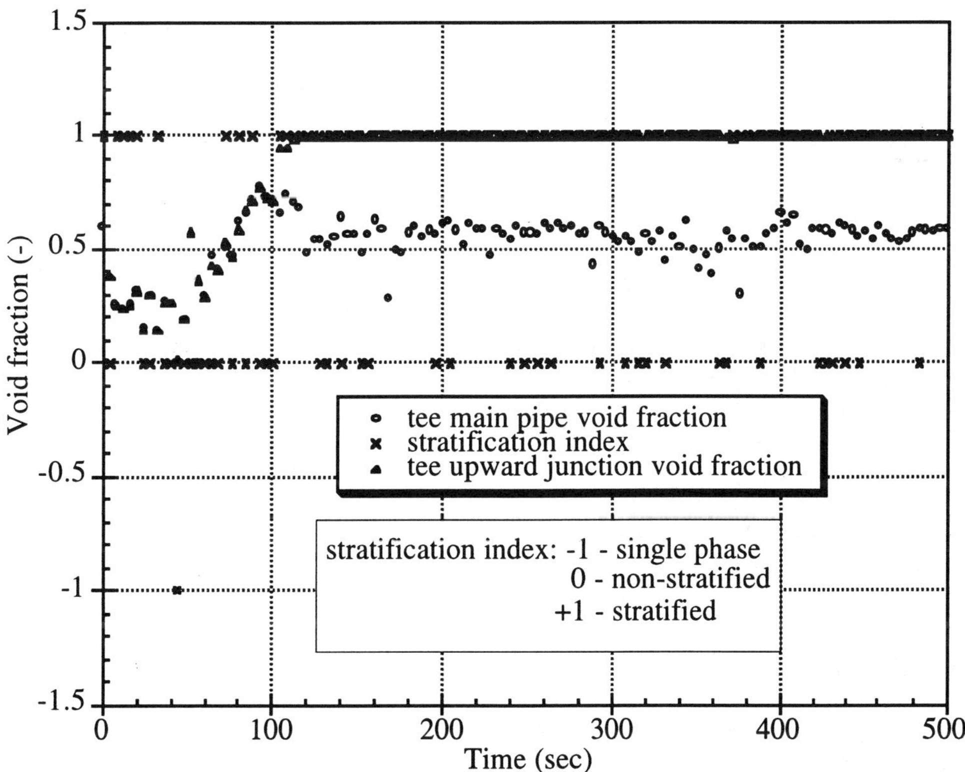

Fig. 13. Void fraction at the tee junctions and HL1 stratification index (CATHARE).

6. Conclusions

The goal of this study was to verify the ability of thermal hydraulic CATHARE V1.3U code to simulate ISP38 loss of RHRS experiment conducted at BETHSY Integral Test Facility. Overall, the code's prediction and experimental data were found to be in a reasonable qualitative agreement. However, calculated timing of the main events during this transient was underestimated.

An investigation to identify the cause of this deficiency showed that the overall system behavior during experiment was strongly affected by the phase separation at surge line/HL1 tee junction and liquid entrainment into the surge line. The code's failure to calculate accurately the flow regime in the HL1 and water entrainment into the surge line led to significant miscalculation of the main events' timing.

Based on the comparison of experimental and calculated results, the following is concluded:

1) the stratification criterion in the horizontal pipe based on the turbulence limitation was experimentally validated for pressure range 2-10 bar and may not be valid for lower pressure as shown in the present experiment;

2) implementing the donor-cell approach in the main pipe and upward tee junction in the case of non-stratified flow conditions which is valid for high pressure may not be valid for low pressure cases and additional experiments are needed to investigate this phenomenon during the lower pressure ranges.

Acknowlegments

Authors would like to express their warmest gratitude to the BETHSY Integral Test Facility and CATHARE teams. Our special thanks go to Dr. G. Lavialle (CEA, France) for his comments and fruitful discussions.

References

D. Dumont, G. Lavialle, B. Noel and R. Deruaz, Loss of residual heat removal during midloop operation: BETHSY experiment, Proceedings, 6-th International Topical Meeting on Nuclear Reactor Thermal Hydraulics, Grenoble, France, 1993, volume 2, 1368-1376.

M. Farvacque, C. Sarrette, CATHARE2 V1.3E: Dictionary of operators and directives, STR/LML-EM/92-124, 1992.

A. Janicot, Qualification report: phase separation in a tee junction, STR/LML/EM/92-135, November 1992.

G. Lavialle, G. Kimber, C. Leveque, BETHSY, International standard problem 38: Loss of residual heat removal system during midloop operation with pressurizer and SG1 outlet plenum manways open, STR/LES/95-244, 1995.

LARGE EDDY SIMULATION OF TUBE BUNDLE GEOMETRIES USING THE DYNAMIC SUBGRID SCALE MODEL

Hagop R. Barsamian
Yassin A. Hassan
Texas A&M University
Department of Nuclear Engineering
College Station, Texas 77843-3133
Phone: (409) 845-7090
Fax: (409) 845-6443

ABSTRACT

The dynamic subgrid scale closure model of Germano et. al (1991) is used in the large eddy simulation computer program, GUST, for incompressible isothermal flows. The advantage of the dynamic subgrid scale model is the exclusion of a model coefficient. This model coefficient is evaluated dynamically at each nodal location for a given timestep by filtering the governing equations on a grid filter and a test filter. A tube bundle geometry of non-staggered arrangement is considered in doubly periodic boundary conditions for two-dimensional simulation. Results of the dynamic subgrid scale simulation are obtained in the form of power spectral densities and flow visualization of turbulent characteristics. Comparisons are performed among the dynamic subgrid scale model, the Smagorinsky eddy viscosity model (Smagorinsky, 1963) using simulations performed with GUST (the Smagorinsky model is used as the base model for the dynamic subgrid scale formulation) and available experimental data. The dynamic subgrid scale model simulations are found to be in good agreement to spectral data available form experiments in similar bundle arrangements. Satisfactory turbulence characteristics are observed through flow visualization of the models. Integral length and time scales are evaluated using correlation curves.

NOMENCLATURE

C	=	model coefficient
D	=	tube diameter
f	=	frequency
L_{ij}	=	resolved turbulent stresses
P	=	pressure
S_{ij}	=	strain rate tensor
S_t	=	Strouhal number
t	=	time
T_{ij}	=	subtest scale Reynolds stress
$\overline{U}$	=	average velocity
$\overline{u}_i$	=	grid filtered flow variable
$\hat{\overline{u}}_i$	=	test filtered flow variable
δ_{ij}	=	kronecker delta
Δ	=	mesh spacing, filter width
τ_{ij}	=	subgrid scale Reynolds stress

INTRODUCTION

The introduction of dynamic closure models has made significant contributions to the application of numerical solutions to the problem of turbulence using the large eddy simulation (LES) methodology. Although turbulence has been observed, studied and analyzed for over 100 years, it is still an unsolved problem. LES in turbulence calculates the large scale characteristics, and models small scale phenomena. In LES, the large scale motion is geometry dependent and is calculated directly through the time integration of the Navier-Stokes equations. The small scales are more universal in nature and are modeled using subgrid scale (SGS) models.

The goal of SGS modeling is to express the fluctuating components of the filtered Navier-Stokes equations in terms of the mean components. Smagorinsky's eddy viscosity model (Smagorinsky, 1963) has been the most widely used SGS model, where the SGS terms are assumed to be proportional to the local mean velocity gradients. Improvements to the Smagorinsky model have persisted, such as the use of damping functions (Moin and Kim, 1982) to account the near-wall effects. In order to overcome the specification of a model constant and better prediction of SGS stresses, Germano et. al (1991) have introduced the dynamic subgrid scale model (DSGS).

In the DSGS model, instead of having the model coefficient as an input parameter, the model coefficient is evaluated dynamically by the introduction of a test filter on top of the grid filter (Fig. 1). The only input parameter is the test to grid filter ratio (recommended value of 2.0 by Germano et. al (1991)). The dynamic evaluation of the model coefficient is preferred, since in complex flows SGS stresses have to be adjusted according to local flow dynamics, giving correct asymptotic behavior near a wall and in laminar flows. Some averaging of the DSGS model coefficient is needed to avoid large variations that may lead to instabilities, and equality is assumed between the grid filter model coefficient and the test filter model coefficient. However, the DSGS model does not rule out backscatter effects (since negative values of the model coefficient are possible) (Germano et. al, 1991).

So far, the DSGS model has been applied to simple flow geometries. Channel flow applications with surface mounted two-dimensional obstacles have been performed by Yang and Ferziger (1993), as well as high-speed axisymmetric transitional boundary layer simulations have been performed by El-Hady et. al (1994) with success. These however, have not been extended to practical applications in complex geometries. One area of practical applications concerns the turbulent buffeting forces experienced by the tubes in steam generators of nuclear power plants. The LES methodology in heat exchanger bundle geometries has been used previously by Stuhmiller et. al (1988) and Pruitt (1992) using only the Smagorinsky SGS model. Barsamian and Hassan (1994) used a modified SGS model that included the cross terms and Leonard term in its formulation, however, this approach has been abandoned due to the Galaleian invariance of the Leonard and cross terms (Piomelli, 1994). Spectral results of these simulations were in reasonable agreement with available experimental data.

The objective of the following investigation is the application of the DSGS closure model using the LES turbulence prediction method in complex flow simulations, and comparison of this to LES results obtained using the Smagorinsky SGS eddy viscosity model and available experimental data. The next section discusses the formulation of the Germano et. al (1991) closure model.

METHODOLOGY

The filtered form of continuity and momentum equations are used in LES. For incompressible, isothermal, Newtonian fluids, these are, respectively:

$$\frac{\partial \overline{u}_i}{\partial x_i} = 0 \tag{1}$$

$$\frac{\partial \overline{u}_i}{\partial t} + \frac{\partial \overline{u}_i \overline{u}_j}{\partial x_j} = -\frac{1}{\rho}\frac{\partial P}{\partial x_i} + v\frac{\partial^2 \overline{u}_i}{\partial x_j \partial x_j} - \frac{\partial \tau_{ij}}{\partial x_j} \tag{2}$$

These filtered equations govern the motion of the large scales. Small scale motions are represented by the SGS stress term given by $\tau_{ij} = \overline{u_i u_j} - \overline{u}_i \overline{u}_j$ that must be modeled.

In the DSGS model, the spatial filtering of the dependent variables leads to two sets of motion equations. These in turn include subgrid scale stresses and the subtest scale stresses given by, respectively (the overbar indicates grid filtered variable, the tophat indicates test filtered variable):

$$\tau_{ij} = \overline{u_i u_j} - \overline{u}_i \overline{u}_j \tag{3}$$

$$T_{ij} = \overline{\widehat{u_i u_j}} - \widehat{\overline{u}}_i \widehat{\overline{u}}_j \tag{4}$$

The resolved turbulent stresses are defined as (where all the terms on the right hand side may be explicitly evaluated) (Germano et. al, 1991):

$$L_{ij} = \overline{\overline{u}_i \overline{u}_j} - \widehat{\overline{u}}_i \widehat{\overline{u}}_j \tag{5}$$

that are related by.

$$L_{ij} = T_{ij} - \hat{\tau}_{ij} \tag{6}$$

Assuming that the same functional form can be used to parameterize both the SGS Reynolds stresses (the Smagorinsky model for this case), the anisotropic part of the grid and test SGS stresses will become, respectively:

$$\tau_{ij} - \frac{1}{3}\delta_{ij}\tau_{kk} = -2C\overline{\Delta}^2 |\overline{S}_{ij}| \overline{S}_{ij} \tag{7}$$

$$T_{ij} - \frac{1}{3}\delta_{ij}T_{kk} = -2C\widehat{\overline{\Delta}}^2 |\widehat{\overline{S}}_{ij}| \widehat{\overline{S}}_{ij} \tag{8}$$

where.

$$|\overline{S}_{ij}| = \left(2\overline{S}_{ij}\overline{S}_{ij}\right)^{1/2} \tag{9}$$

$$\overline{S}_{ij} = \frac{1}{2}\left(\frac{\partial \overline{u}_i}{\partial x_j} + \frac{\partial \overline{u}_j}{\partial x_i}\right) \tag{10}$$

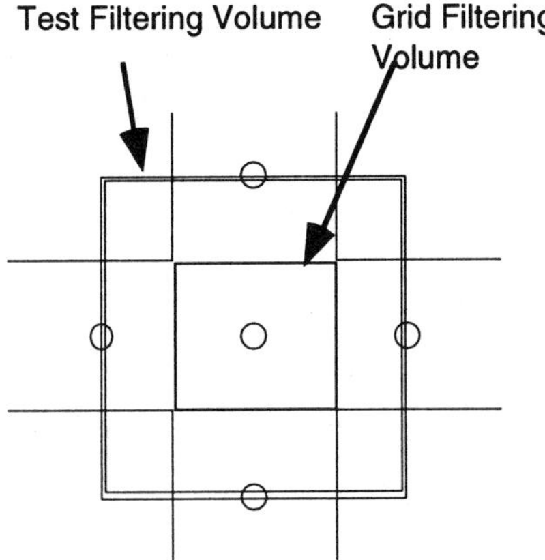

Figure 1. Grid and test filters used in the Dynamic Subgrid Scale model.

Equating the resolved turbulent stresses (Eq. 6) and the modeled SGS stresses (Eqs. 7 and 8), we obtain a value for the model coefficient, C (using the least squares approach to minimize the error locally) (Lilly, 1992):

$$C = -\frac{1}{2}\frac{L_{ij}M_{ij}}{M_{ij}M_{ij}} \tag{11}$$

where.

$$M_{ij} = \overline{\hat{\Delta}^2\left|\hat{\overline{S}}_{ij}\right|\hat{\overline{S}}_{ij}} - \overline{\Delta^2\left|\overline{S}_{ij}\right|\overline{S}_{ij}} \tag{12}$$

The above method produces a computationally more stable value of C than Germano et. al's method. The only parameter needed to be specified in the DSGS model is the ratio between the test filter width, $\hat{\overline{\Delta}}$, and grid scale filter width, $\overline{\Delta}$. Germano et. al (1991) have shown that LES results are not overly sensitive to this ratio, and in this study $\hat{\overline{\Delta}}/\overline{\Delta} = 2.0$ was used. Unlike Germano et. al's method that uses contraction, the denominator in Lilly's formulation vanishes only when all nine components vanish at the same time.

COMPUTATIONAL TECHNIQUE

The calculations in this study are performed using the GUST computer program (Chilukuri, 1987; Lee, 1992). GUST is a FORTRAN computer program employing the LES technique for incompressible isothermal flow turbulence in complex, three-dimensional geometries. Complex geometries are represented in GUST using partially blocked cells. For the no-slip boundary condition imposed on solid walls the LES code GUST assumes a linear relationship between the velocity adjacent to a wall and the shear stress.

GUST uses a Factorized Implicit Solution Technique (FIST) in solving the space-filtered Navier-Stokes equations. FIST is an efficient finite difference algorithm to advance a multidimensional solution field in time, without having to solve large systems of coupled algebraic equations at each timestep.

The procedure for advancing the solution in time involves a predictor-corrector scheme (similar to that of the SMAC method). The mass fluxes (used as the dependent variables in GUST) at each new timestep are determined by integration of the momentum equations (given a pressure field for the new timestep and suitable linearization of the advection terms). However, in general, the pressure filed at the new timestep is not known a priori. Hence, a best available estimate for pressure is used to advance the mass fluxes to the new timestep. At this point though, mass conservation is not satisfied, therefore, iteration is performed within the given timestep until mass and momentum conservation are satisfied.

GUST has been tested and applied to practical flow situations with relative success. Three-dimensional channel flow cases have been performed with GUST and compared to experimental data (Chilukuri, 1987). Testing and fine tuning of the Smagorinsky model coefficient has been performed in channel flow and tube bundle simulations with relatively sparse nodalization schemes (Stuhmiller et. al, 1988). Practical applications of rigid single cylinder simulations have been performed by Hassan and Bagwell (1990) and Hassan and Lee (1993), where good agreement was shown between numerical and experimental spectral densities. However, the size of the computational domain was not sufficiently large to correctly account for the long wave fluctuations. Experimental evidence indicates that the turbulent memory spans about three to four tube pitches in the mean flow direction (Chen and Jendrzejczyk, 1987). Pruitt et. al (1992) have performed tube bundle simulations for the entrance region of the Westinghouse D4 steam generator with favorable results of the drag and lift forces.

RESULTS AND DISCUSSIONS

In order to study the influence of grid size, a single tube case (using the Smagorinsky closure model) was considered. Arbitrary pitch-to-diameter ratio (p/d) and tube diameter were chosen, 0.04445 m and 0.0254 m, respectively. Grid refinement shows that the spatial resolution is adequate beyond the 40 x 40 sized case. Using denser nodalizations shows improvement, but the difference is insignificant. Grid independence is declared beyond the 50 x 50 case. Subsequent simulations were performed with 50 meshes per pitch length in bundle geometries. The resulting Reynolds number produced a $\Delta x^+ = \Delta y^+ = 140$.

Spatial averaging of C has been suggested by Germano et. al, however, it is difficult to do so in complex geometries due to the lack of homogeneous flow directions. Large variations

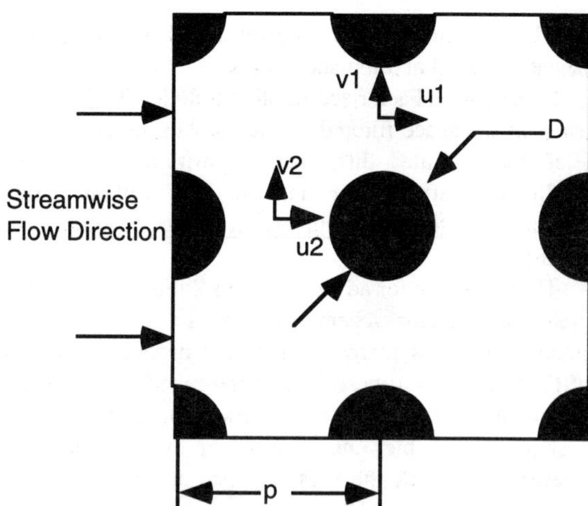

Figure 2. Non-staggered bundle geometry.

of C may lead to instabilities, to overcome this problem, averaging of the dynamically evaluated model coefficient is performed in space. This was done by averaging the value of C using neighboring points for each locations at each timestep. Upper and lower limits were also set so as to avoid large variations of C. The simulations were performed on SUN-Sparc workstations and SGI Power Challenge supercomputer.

The test simulation involved a non-staggered tube bundle array with tube diameter of 2.54 cm and p/d = 1.75. The bundle simulation was deemed necessary due to correlation lengths that are on the order of four tube diameter lengths and were not satisfied by the single tube simulation. This is somewhat of a rather arbitrary scaling factor, since not much experimental data is available on the matter. Using 50 cells per pitch length, a 100 x 100 mesh geometry (with a total of 10,000 computational cells) was generated including nine tube centers. Figure 2 gives a sketch of the geometry with pertinent information. Doubly periodic boundary conditions were applied with a pressure gradient, $-\dfrac{1}{\rho}\dfrac{\partial p}{\partial x}$, of 100 N/Kg used as the driving force in the streamwise direction. The simulations were performed using a timestep of 0.2 msec for 3000 timesteps. The average gap velocity corresponded to a Reynolds number of 1.65×10^5 based on tube diameter.

Visualization of the simulations utilizing both closure models provided an insight into vortex shedding excitation. Vortex shedding excitation is caused by pressure fluctuations due to their periodical nature. These forces occur mainly in the lift direction when the vortex shedding frequency matches the natural frequency of a tube and large amplitude vibrations may occur. However, it is generally accepted (Paidoussis, 1982; Weaver and Grover, 1978; Owen, 1965) that vortex shedding has no practical significance inside closely packed tube bundles as will be seen by the absence of a narrow band peak in the lift directions power spectral density (PSD) plots.

Figure 3 gives the fluctuating pressure contours for both closure model simulations at timestep 2500. The observed

pressure fluctuations at the tube walls are maximum at about 90 degrees from the mean flow direction as has been observed experimentally. Figure 4 gives the streamlines of the bundle simulation for both closure models at timestep 2500. These show the switching phenomena that has been observed experimentally (Moretti, 1993). The switching changes the mean direction of flow in the horizontal gap between two adjacent tubes, as the mean flow continues in the direction of applied pressure gradient. Figure 5 delinates the velocity vectors at timestep 2500 for each closure model. In both simulations, acceptable characteristics are observed.

Experimental data for fluctuating quantities in non-staggered tube bundle geometries is sparse. Chen and Jendrzejczyk (1987) examined the fluctuating forces on a non-staggered bundle with a p/d = 1.75. The experimental setup was for water flow past a 7 x 7 array of one inch diameter tubes. Upstream turbulence was generated by an upstream grid. Three different grids were used for comparison. The fluctuating force on the tubes was measured using piezoelectric transducers. PSD plots for the fluctuating lift and drag forces were available only at the second row and last row of tube cells. Additionally, due to experimental constraints, they were unable to calculate the mean values for the lift and drag coefficients. The force PSD plots for the second tube row show a significant contribution from the turbulence associated with the type grid that is immediately before the tube region, therefore, it was excluded from the comparison. Although the seventh tube row is the last tube row and hence may have some end effects associated with a large degree of vortex shedding, it will be a better comparison to the computed data, due to the fact that it is independent of the effects of the grids.

Using the available experimental data, normalized PSDs of lift and drag forces of both closure models (Smagorinsky and DSGS models) are compared as a function of the Strouhal number, S_t, defined by:

$$S_t = \frac{fD}{\overline{U}} \tag{13}$$

where, f is the corresponding frequency, D is tube diameter, and $\overline{U}$ is the average gap velocity. The PSDs have been defined as the square of the Fourier transform of the lift and drag forces.

Figure 6 shows the normalized PSD of the drag force as a function of Strouhal number. Here, the DSGS results are in good agreement with that of the experimental data. The dissipation rate of the DSGS model at the higher frequencies agrees well with that of experimental data as well. Figure 7 gives results for the normalized lift force. As mentioned above, there is no pronounced narrow band peak defined indicating that vortex shedding excitation is not prominent, as expected (Paidoussis, 1982; Weaver and Grover, 1978; Owen, 1965). There is a shift in the Strouhal number position between the numerical and experimental findings. This may be

Table 1. Integral time and length scales.

	Smagorinsky Model	Dynamic SGS Model
Integral Time	(sec)	(sec)
u1	1.34e-2	9.27e-3
u2	2.46e-3	1.95e-3
v1	2.34e-3	1.52e-3
v2	3.39e-3	3.05e-3
Integral Length	(m)	(m)
u1	1.02e-1	8.71e-2
u2	7.56e-4	1.14e-3
v1	4.76e-4	5.19e-4
v2	2.23e-4	6.59e-4

due to the differences in Reynolds number of the simulations, or three-dimensional effects that are not captured in two-dimensional simulations.

Figure 8 contains the normalized autocorrelation function of the streamwise velocities with the locations as indicated in Fig. 2. As shown in Table 1, the integral time and length (using Taylor's frozen field hypothesis) scales of the streamwise velocity at location 1 are much larger than that of location 2, for both closure model simulations. Figure 9 contains the normalized autocorrelation function of the transverse velocities. The autocorrelations display the conventional decay function, where there is initial rapid decay followed by a more gradual decay.

CONCLUSIONS

The dynamic subgrid scale model has been successfully used in complex flow situations. A non-staggered deep bundle simulation is performed using the dynamic subgrid scale model and the Smagorinsky eddy viscosity model. Numerical results are compared to that of experimental data in the form of power spectral densities. It is observed that the dynamic subgrid scale spectra correlate well with that of experimental data. Therefore, the dynamic subgrid scale model may be considered as a closure model to turbulence simulations for the large eddy simulation method in complex geometries. It should be noted that using the Smagorinsky model as a base model to the dynamics subgrid scale model limits the accuracy of the dynamic model due to the assumptions made in the Smagorinsky model. Therefore, the use of a better base model is recommended. Further detailed comparison to direct numerical simulation or experimental data is deemed necessary when data of this kind becomes available for complex geometries.

ACKNOWLEDGMENTS

The authors gratefully acknowledge the considerable guidance, participation and insights of Dr. G. Srikantiah, EPRI Project Manager, and Dr. D. A. Steininger throughout the course of this work.

REFERENCES

Barsamian, H. R., and Hassan, Y. A., 1994, "Modified Subgrid Scale Model for Large Eddy Simulation of Tube Bundle Cross Flows," *Proc. Sym. on Flow-Induced Vibration*, Vol. 273, pp. 283-288.

Chen, S. S., and Jendrzejczyk, J. A., 1987, "Fluid Excitation Forces Acting on a Tube Array," Argonne National Laboratory Report ANL-85-55, Chicago, Illinois.

Chilukuri, R., 1987, "GUST: A Computer Program for Large Eddy Simulation of Incompressible Isothermal Flow Turbulence," *Theory Manual*, EPRI Project S309-2, Palo Alto, CA.

El-Hady, N. M., Zang, T. A., and Piomelli, U., 1994, "Application of the Dynamic Subgrid-Scale Model to Axisymmetric Transitional Boundary Layer at High Speed," *Physics of Fluids A*, Vol. 6(3), pp. 1299-1309.

Germano, M., Piomelli, U., Moin, P., and Cabot, W. H., 1991, "A Dynamic Subgrid-Scale Eddy Viscosity Model," *Physics of Fluids A*, Vol. 3(7), pp. 1760-1765.

Hassan, Y. A., and Bagwell, T. G., 1990, "Large Eddy Simulation on Supercomputers," *Project Report*, EPRI-NP-7041, Palo Alto, CA.

Hassan, Y. A., and Lee, S. Y., 1993, "Application of Large Eddy Simulation to Three-Dimensional Tube Bundle Flows," *NURETH-6*, Vol. 2, pp. 1415-1419.

Lee, S. Y., 1992, "A Study and Improvement of Large Eddy Simulation (LES) for Practical Application," Ph.D. Dissertation, Texas A&M University, College Station, Texas.

Lilly, D. K., 1992, "A Proposed Modification of the Germano Subgrid-Scale Closure Method," *Physics of Fluids A*, Vol. 4(3), pp. 633-635.

Moin, P., and Kim, J., 1982, "Numerical Investigation of Turbulent Channel Flow," *Journal of Fluid Mechanics*, Vol. 118, pp. 341-377.

Moretti, P. M., 1993, "Flow-Induced Vibrations in Arrays of Cylinders," *Annu. Rev. Fluid Mechanics*, Vol. 25, pp. 99-114.

Owen, P. R., 1965, "Buffeting Excitation of Boiler Tube Vibration," *Journal of Mechanical Engineering Science*, Vol. 7, pp. 431-439.

Paidoussis, M. P., 1982, "A Review of Flow Induced Vibration in Reactors and Reactor Components," *Nuclear Engineering and Design*, Vol. 74, pp. 31-60.

Piomelli, U., 1994, "Large-eddy Simulation of Turbulent Flows," Theoretical and Applied Mechanics Report No. 767, UILU-ENG-94-6023, University of Illinois at Urbana-Champaign, Illinois.

Pruitt, J. M., 1992, "Large Eddy Simulation of Turbulence within Heat Exchangers," Master's Thesis, Texas A&M University, College Station, Texas.

Pruitt, J. M., Hassan, Y. A., and Steininger, D. A., 1992, "Large Eddy Simulation of Turbulence Entering a PWR Steam

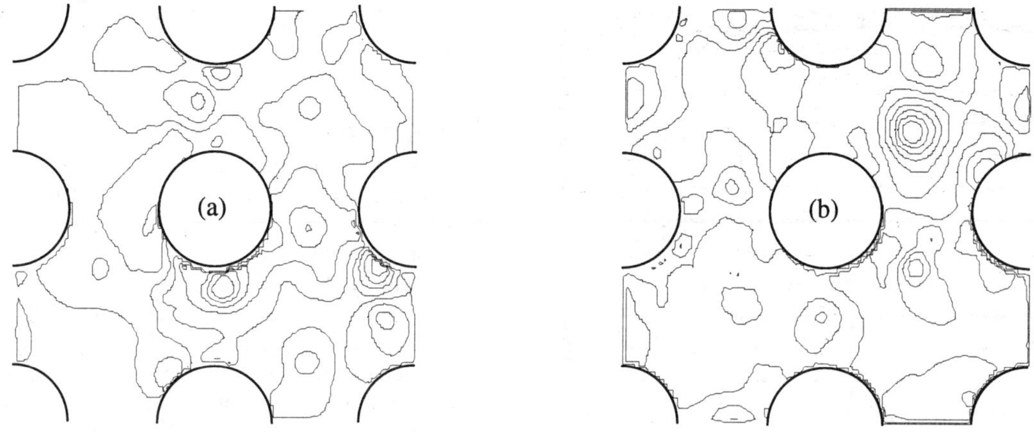

Figure 3. Pressure contours at timestep 2500, for (a)Smagorinsky SGS Model, (b)Dynamic SGS model.

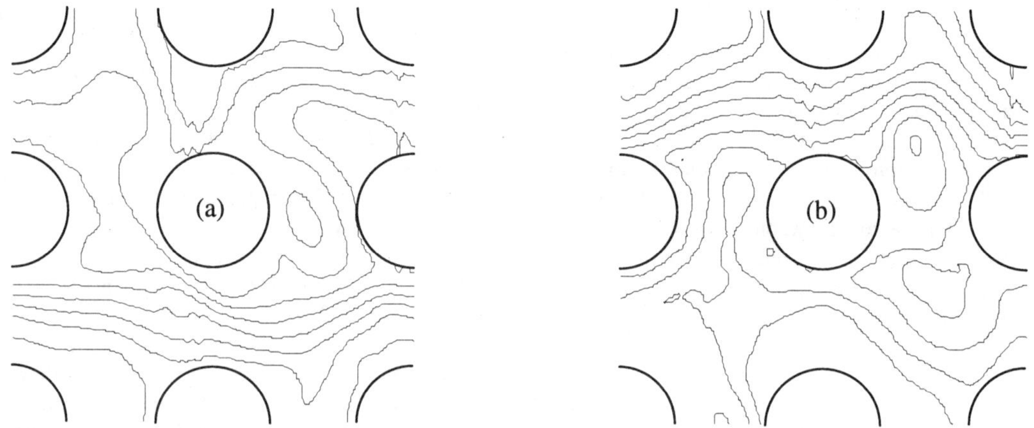

Figure 4. Streamlines at timestep 2500, for (a)Smagorinsky SGS Model, (b)Dynamic SGS model.

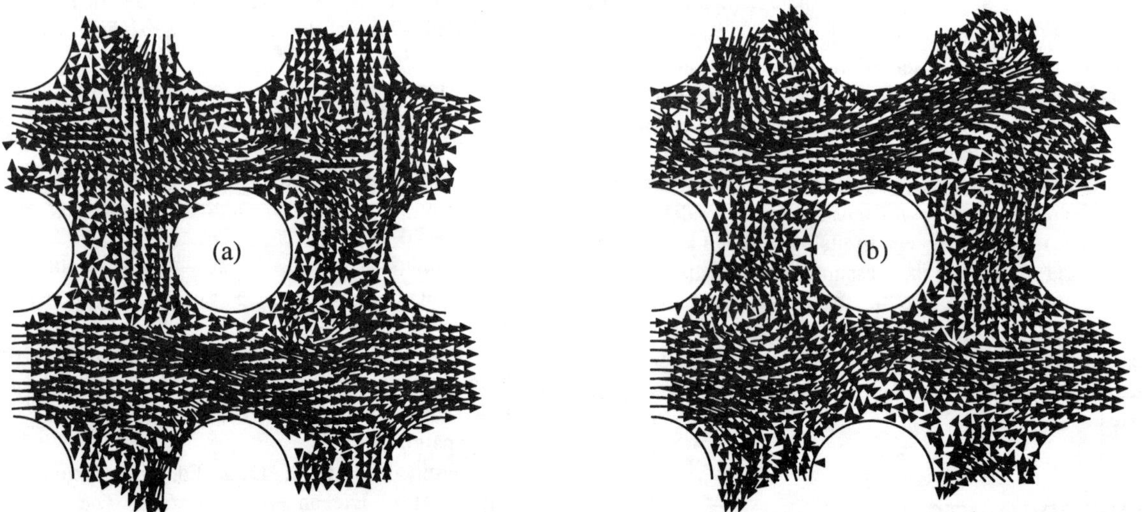

Figure 5. Velocity vectors at timestep 2500, for (a)Smagorinsky SGS Model, (b)Dynamic SGS model.

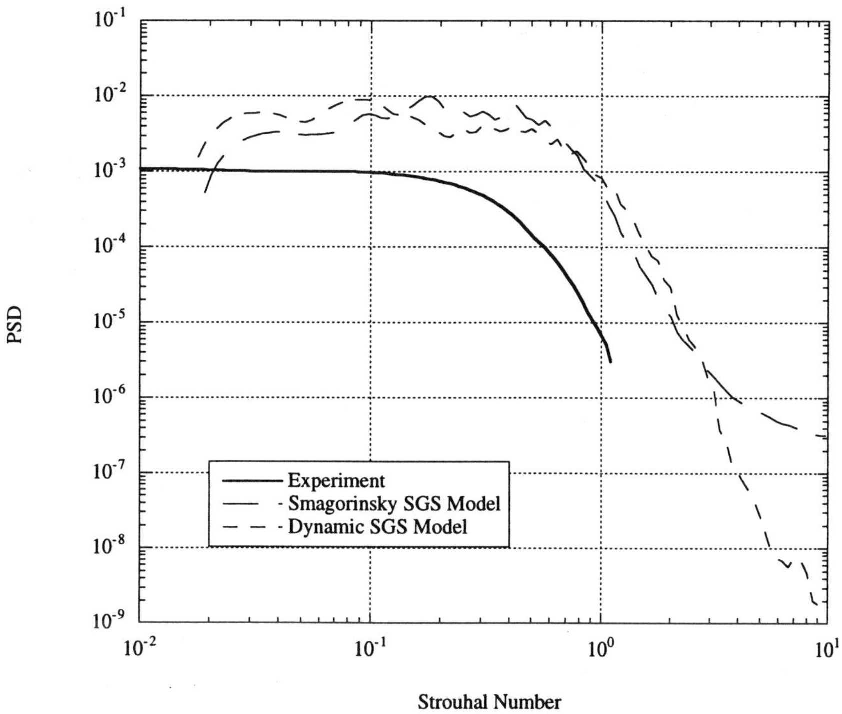

Figure 6. Drag force PSD as a function of Strouhal number.

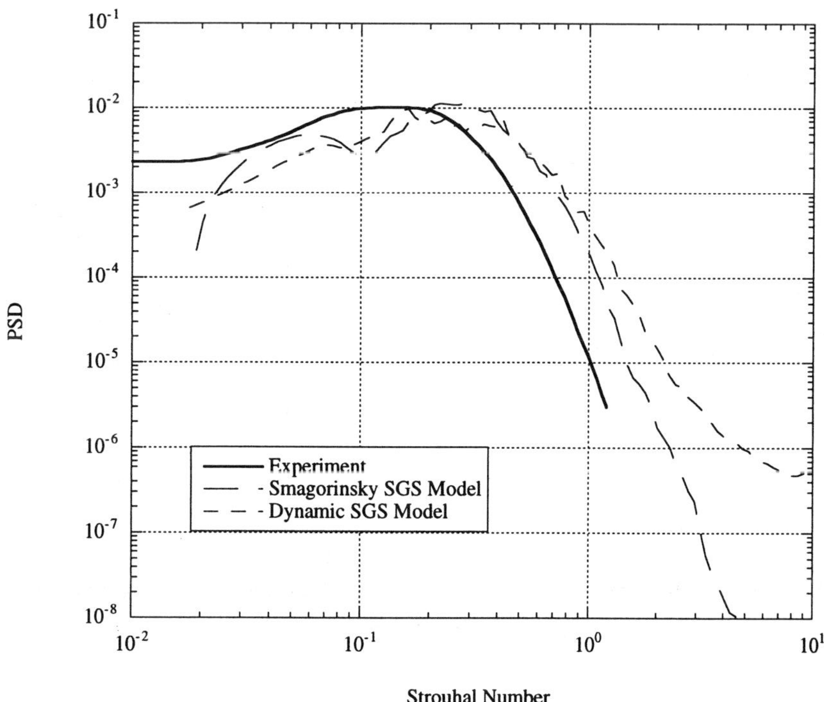

Figure 7. Lift force PSD as a function of Strouhal number.

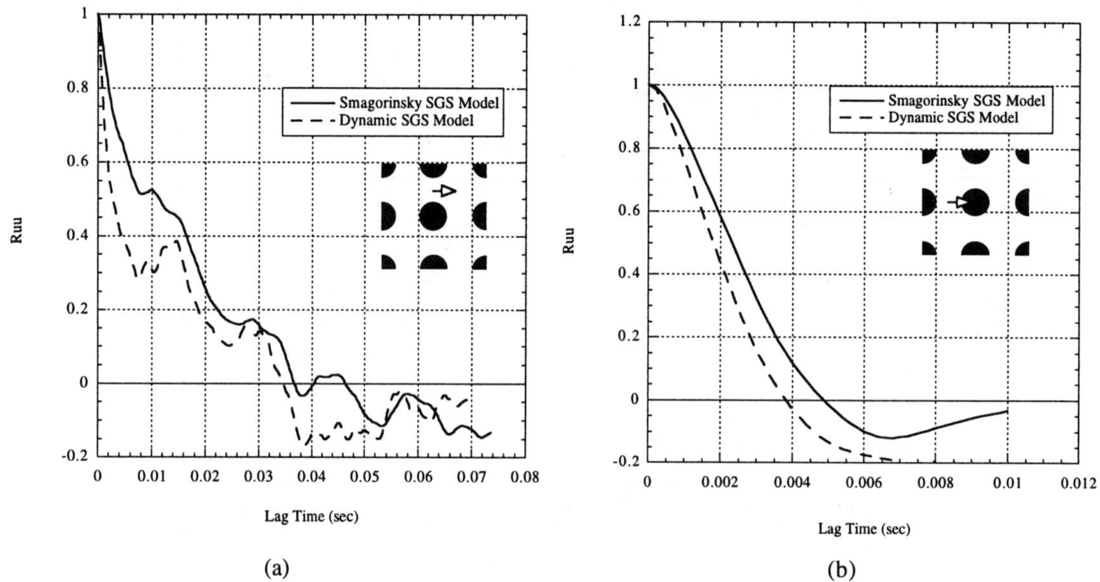

**Figure 8. Normalized autocorrelation of streamwise velocities at
(a) location 1, (b) location 2.**

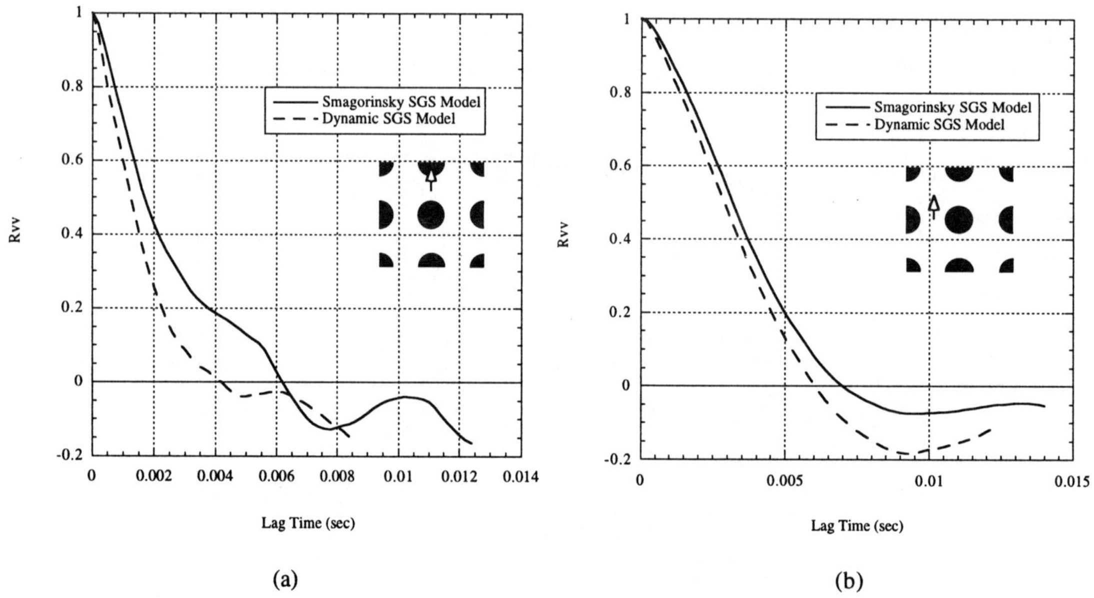

**Figure 9. Normalized autocorrelation of transverse velocities at
(a) location 1, (b) location 2.**

Generator," *Proc. Sym. on Flow-Induced Vibration and Noise*, Vol. 7 AMD-151 PVP-247, pp. 173-183.

Smagorinsky, J., 1963, "General Circulation experiments with Primitive Equations," *Monthly Weather Review*, Vol. 91, pp. 216-241.

Stuhmiller, J. H., Chilukuri, R., Masiello, P. J., Chan, R. K.-C., Yu, J. H.-Y., and Ho, K. H.-H., 1988, "Prediction of Localized Flow Velocities and Turbulence in a PWR Steam Generator," *Project Report*, EPRI-NP-5555, Palo Alto, CA.

Weaver, D. S., and L. K. Grover, 1978, "Cross-Flow Induced Vibrations in a Tube Bank- Turbulent Buffeting and Fluid Elastic Instability," *J. of Sound and Vibration* , Vol. 59, pp. 277-294.

Yang, K. S., and Ferziger, J. H., 1993, "Large-Eddy Simulation of Turbulent Flow in a Channel with a Two-Dimensional Obstacle Using a Dynamic Subgrid-Scale Model," *AIAA-93-0542*, pp. 1-8.

TURBULENT FRICTION FACTOR FOR TWO-PHASE: AIR-VISCOELASTIC FLUID FLOWS THROUGH HORIZONTAL STRAIGHT CIRCULAR TUBES

B. K. Rao, College of Engineering, Idaho State University, Pocatello, Idaho 83209

ABSTRACT

Fully developed turbulent isothermal Fanning friction factors for air-viscoelastic fluid flows through a horizontal pipe were experimentally measured. The viscoelastic fluids studied were aqueous solutions of polyacrylamide at three different concentrations (100, 200, and 500 wppm). The two-phase friction factor data were correlated in terms of the apparent Reynolds number defined based on the effective viscosity for gas-liquid mixture. Over a range of the apparent Reynolds number from 10^4 to 10^5, the simpler homogeneous model used was found to be accurate enough for engineering prediction of friction factor for air-viscoelastic fluid flows through straight pipes. A correlation for the prediction of air-viscoelastic two-phase turbulent friction factor is proposed.

NOMENCLATURE

A_c cross-sectional area of tube [m^2] = π D^2/4

D   tube inside diameter [m]

K'   consistency index of power-law fluid

L   distance between pressure taps [m]

$\dot{m}$ mass flow rate [kg/s]

$\dot{Q}$ volume flow rate = $\dot{m}$ / ρ [m^3/s]

v   mean velocity = $\dot{Q}$ / A_c [m/s]

v_{TP} velocity for 2-phase = $\dot{m}_t$ / (A_c ρ_{TP})

z   axial distance [m]

μ viscosity [Pa.s]

η apparent viscosity of polymer solution [Pa.s]

ρ density of fluid [kg/m^3]

λ characteristic time of viscoelastic fluid [s]

τ_w wall shear stress = Δp D / (4L) [Pa]

Δp   pressure drop, 2 f L v^2 ρ / D [Pa]

Dimensionless quantities

ϵ volumetric quality = $\dot{Q}_g$ / ($\dot{Q}_g$ + $\dot{Q}_l$)

ϕ^2 frictional multiplier =

Δp_{TP} / Δp_{sp}

f   Fanning friction factor = τ_w / ($\frac{1}{2}$ ρ v^2)

F   Froude number = $[(\rho_g / (\rho_1 - \rho_g)) / (D\ g)]^{1/2}\ v_g$

K^2 $F^2\ Re_1$

n   power-law exponent

Re   Renolds number = $\rho\ v\ D$ / μ = $4\ \dot{m}$ / $(\pi\ D\ \mu)$

Re_a apparent Reynolds number = $\rho\ v\ D$ / η

T $\{(dp/dz)_1 / [(\rho_1 - \rho_g)g]\}^{1/2}$

x   mass quality = $\dot{m}_g$ / $(\dot{m}_g + \dot{m}_1)$

X^2 Δp_1 / Δp_g

Ws   Weissenberg number = $\lambda\ v$ / D

Suffixes:

m   measured
p   predicted
sp   single phase (g: gas; l: liquid; t: total)
TP   two-phase
w   at the wall

INTRODUCTION

Applications of two-phase gas-liquid flows range from transfer systems, such as pumping, to those involving heat and/or mass transfer, such as heat exchangers, thermosyphon reboilers, scrubbers, cooling towers, bubble columns and nuclear reactors.

The immense volume of literature on gas-liquid flows deals with two-phase flows involving gases and Newtonian liquids and has often produced confusion for the designer because of arbitrary definitions, flow orientations, and their combinations. The designer's key interests in adiabatic flows are in the following parameters: void fraction, and frictional pressure drop. These parameters are related to various controlled variables such as fluid properties, flowrates, and the size, configuration and orientation of equipment.

Of the four types of two-phase flow (gas-liquid, gas-solid, liquid-liquid, and solid-liquid), gas-liquid flows are the most complex, since one of the phases is compressible. For given flows of the two phases in a given channel, the gas-liquid interfacial distribution can take several possible forms resulting in various *flow patterns or flow regimes* (eg. bubble flow, annular flow etc).

For pressure drop calculations, the invocation of flow patterns is arguably unnecessary. However, the relationships for pressure drop are likely to be sensitive to the flow-patterns. Hence, the understanding and delineation of flow patterns is very important. Assessment of the flow regime in any given situation is somewhat subjective. A wide variety of methods have been proposed in the literature for determination of flow regimes (Fiori and Bergles 1966, Dsarasov et al 1974, Hewitt

1978, Reimann and John 1978, and Kutateladze 1973).

The nature of the flow regimes varies with channel geometry and orientation. Flow regimes in horizontal flow are more complex than those in vertical flow (due to the asymmetry in the flow induced by the gravitational force acting normal to the direction of flow. The preponderance of data on flow patterns in horizontal tubes are for air-water flows (typically in tubes of 2-5 cm diameter). There is a dearth of horizontal flow data covering a wide range of physical properties.

Sakaguchi et al (1979) classified the flow regimes based on superficial velocities of liquid and gas phases. One of the best known generalized-flow-pattern maps for horizontal flow is that of Baker (1954). Baker took account of physical properties by introducing the following parameters:

$$\lambda = [(\rho_g/\rho_a)(\rho_l/\rho_w)]^{1/2} \quad (1)$$

$$\Psi = (\sigma_w/\sigma)[(\mu_l/\mu_w)(\rho_w/\rho_l)]^{1/3} \quad (2)$$

where ρ, σ, and μ represent, respectively, density, surface tension, and viscosity, the subscripts g and l represent the gas and liquid phases, and the subscripts a and w represent the values for air and water at 20°C and atmospheric presure. Scott (1963) modified the original form of the Baker's map, introducing the parameters $(\dot{m}_l\Psi\lambda/\dot{m}_g)$ and $\dot{m}_g/\lambda$; this revised map has the advantages of showing transition regions indicating uncertainty between the various regimes. Bell et al (1969) and Collier (1972) used $\dot{m}_g/\lambda$ and $\dot{m}_l\Psi$ as the coordinates for their flow regime maps.

To represent the effects of various system parameters, several authors have developed a number of alternative flow maps (Schicht 1969, Al-Sheikh et al. 1970, Mandhane et al. 1974, Fisher and Yu 1975). However, none of these maps represents all the appropriate *transitions* in terms of a single set of parameters.

Taitel and Dukler (1976) introduced, based on a semitheoretical approach, dimensionless quantities K, T, and F for extrapolating the correlations to a wider range of pipe sizes and fluid properties. Their generalized-flow-regime map for horizontal two-phase flow is in terms of K, T, F, and X which are defined as follows:

$$F^2 = \{\rho_g/[(\rho_l-\rho_g)Dg]\} (U_g)^2 \quad (3)$$

$$K^2 = F^2 Re_l \quad (4)$$

$$Re_l = DU_l/\nu_l \quad (5)$$

$$T^2 = (dp/dz)_l/[(\rho_l-\rho_g)g] \quad (6)$$

$$X^2 = (dp/dz)_l/(dp/dz)_g \quad (7)$$

where U and ν are the superficial velocity and the kinematic viscosity of the fluid respectively. The nature and prediction of the transition are still the subjects of active research. Notable among these studies are those of Wallis and Dobson (1973), Gardner (1977), and Kubie (1979).

Several analytical models are available to describe the two-phase flows: the *homogeneous* model, *separated-flow* model, *drift-flux* model etc. (Wallis, 1969).

Homogeneous Model is based on the assumption that velocities of the two phases are equal (*slip ratio* = 1). This reduces the homogeneous density and void fraction to the following:

$$\epsilon = x/[x+(1-x)\rho_g/\rho_l] \quad (8)$$

$$\rho_{TP} = \epsilon \, \rho_g + (1-\epsilon) \, \rho_l \quad (9)$$

$$\rho_{TP} = \rho_g\rho_l/[x\rho_l+(1-x)\rho_g] \quad (9a)$$

where x is the quality of the mixture based on mass flow rates. The two-phase friction factor (f_{TP}) is defined as a function of the two-phase Reynolds number (Re_{TP}).

$$Re_{TP} = 4 \, \dot{m} \, / \, (\pi \, D \, \mu_{TP}) \quad (10)$$

$$f_{TP} = (\Delta p/dz) \, D^5 \, \pi^2 \, \rho \, / \, (32 \, (\dot{m}^2)) \quad (11)$$

Several definitions for the two-phase viscosity (μ_{TP}) are reported in the literature. The most widely used is that due to McAdams et al. (1942), which is

$$1/\mu_{TP} = x/\mu_g + (1-x)/\mu_l \quad (12)$$

The frictional pressure gradient is related to that for the gas phase or liquid phase flowing alone in the channel, in terms of frictional multipliers ϕ_g and ϕ_l defined as follows:

$$(\phi_g)^2 = (dp/dz)_{TP}/(dp/dz)_g \quad (13)$$

$$(\phi_l)^2 = (dp/dz)_{TP}/(dp/dz)_l \quad (14)$$

The single-phase pressure drops in Eqs (13 and 14) are given by:

$$(dp/dz)_g = 32 \, f_g \, \dot{m}^2 \, x^2/ \, \rho \, \pi^2 \, D^5 \quad (15)$$

$$(dp/dz)_l = 32 \, f_l \, \dot{m}^2 \, (1-x)^2 \, / \, \rho \, \pi^2 \, D^5 \quad (16)$$

The single phase Reynolds numbers Re_g and Re_l are given by:

$$Re_g = 4 \, \dot{m} \, x \, / \, (\pi \, D \, \mu_g) \quad (17)$$

$$Re_l = 4 \, \dot{m} \, (1-x) \, / \, (\pi \, D \, \mu_l) \quad (18)$$

Separated Flow Model is based on the assumption that the two phases can have different velocities. The famous Lockhart & Martinelli (1949) correlation for frictional pressure drop in terms of pressure drop multipliers is:

$$(\phi_l)^2 = 1 + C/X + 1/X^2 \quad (19)$$

$$(\phi_g)^2 = 1 + CX + X^2 \quad (20)$$

where C is a dimensionless parameter whose value depends on the nature of the phase-alone flows. The *drift-flux* model takes into account the relative motion of two phases.

NON-NEWTONIAN FLUIDS

NonNewtonian fluids such as polymer melts, paints, food products, petrochemical and biochemical products are frequently encountered in these process industries. Knowledge of the hydrodynamic behavior of these rheologically complex fluids is essential to an improved design of the equipment handling such fluids.

NonNewtonian fluids exhibit nonlinear relationship between shear stress (τ) and shear rate (γ). The simplest mathematical model that describes the flow behavior of a nonNewtonian fluid is given by

$$\tau = K'\gamma^n \qquad (21)$$

K' and n are rheological constants. The apparent viscosity (η) of nonNewtonian fluids, and the apparent Reynolds number are defined as:

$$\eta = \tau / \gamma \qquad (22)$$
$$Re_a = 4 \dot{m} / \pi D \eta \qquad (23)$$

An introduction to the classification of nonNewtonian fluids, and the constitutive equations for hydrodynamics of these fluids are provided by some excellent references: Bird et al. (1960), Middleman (1968), Skelland (1967) and Wilkinson (1960).

Viscoelastic fluids are a special class of nonNewtonian fluids. These fluids have elasticity and therefore can sustain unequal normal stresses at the flow boundaries and interfaces. The experimentally observed hydrodynamic entrance length for turbulent viscoelastic flows in straight circular pipes is about 150 hydraulic diameters. The turbulent friction factor for aqueous solutions of viscoelastic polymer is lower than that for the Newtonian solvent (Cho and Hartnett 1982). This phenomena is known as *drag reduction*. As the polymer concentration is increased, friction factor monotonically decreases, at any given Reynolds number, upto a minimum value (**the friction factor asymptote**); with further increase in polymer concentration, the solution becomes more viscous and increasingly elastic but, the

friction factor will not be affected. This anomaly has to do with the definition of the apparent Reynolds number. The minimum concentration of polymer needed to achieve the friction factor asymptote depends on the chemistry of the solvent used, the polymer itself, cross-section of the flow passage etc.

The fully developed turbulent friction factor for viscoelastic dilute polymer solutions in straight pipes is correlated in terms of the Weissenberg number, Ws (Cho and Hartnett 1982):

$$Ws = \lambda V / D \qquad (24)$$
where λ is the characteristic time of the viscoelastic fluid, and V is the mean velocity of the fluid.

Information on gas-nonNewtonian liquid flows is scarce. This experimental investigation is undertaken to study the pressure drop and friction factor for two-phase flows (air-viscoelastic fluid) in horizontal tubes.

EXPERIMENTAL SETUP AND PROCEDURE

The schematic of the flow loop is shown in Figure 1. The flow loop consists of a reservoir, a positive displacement pump, an opaque test section (0.025-m diameter and 11 m-long PVC tube), a calibrated magnetic flow meter, a rotameter, and a differential pressure transducer. Only plastic piping and valves were used in the flow loop to minimize any possible chemical reaction between the test fluid and the components of the flow loop.

The viscoelastic fluids studied were aqueous solutions of polyacrylamide. Distilled water was used as a Newtonian solvent. The test fluids were prepared in the reservoir by dissolving polymer powder (of known weight) in distilled water at room temperature, employing only hand stirring (to minimize mechanical degradation). The polymer used to prepare the test fluids was Praestol (produced by Stockhausen, Inc) and came from a single batch. The test fluid concentrations employed were 100, 200, and 500 wppm (parts per million by weight).

Air was injected at about 15 diameters downstream of the tube entrance. The liquid flow rate was varied from zero to 4.2×10^{-3} m^3/s in steps of about 0.42×10^{-3} m^3/s. At each of these liquid flowrates, a set of runs was completed by varying the air flow from zero to 1.2×10^{-3} m^3/s in steps of abput 0.2×10^{-3} m^3/s. For each run, the liquid flow rate, air flow rate, and pressure drop (between taps at 250 and 300 diameters from the entrance) were measured.

The test fluid (polymer solution) steady shear viscosity was measured at room temperature using a Bohlin rheogoniometer, a Brookfield viscometer, and a capillary viscometer. The shear stress (τ) is measured as a function of shear rate ($\dot{\gamma}$), and a parabolic equation is fitted in the range of wall shear stress (τ_w) measured in the test section:

$$ln \; \tau = a_0 + a_1 \; ln \; \dot{\gamma} + a_2 \; (ln \; \dot{\gamma})^2 \tag{25}$$

The power-law exponent (n), consistency index (K'), and apparent viscosity (η) were calculated as follows:

$$n = a_1 + 2 \; a_2 \; (ln \; \dot{\gamma}) \tag{26}$$
$$K' = \exp[a_0 - a_2 \; (ln \; \dot{\gamma})^2] \tag{27}$$
$$\eta = \tau \; / \; \dot{\gamma} \tag{28}$$

The characteristic curves are shown in Figure 2. For each set of runs (with a fixed concentration of polymer), a fresh batch of solution was prepared. The test fluid rheology was measured before and after each run to monitor the polymer degradation, and the variation was within ± 2%.

RESULTS AND DISCUSSION

The author is unaware of information on hydrodynamic entrance length for turbulent two-phase gas-viscoelastic flows. The present data for air-water as well as air-viscoelastic fluids are believed to be obtained in the hydrodynamically fully developed region. It may be noted that the flow-regime maps in the literature were developed for gas-Newtonian liquids. The measured air mass flux values are shown as a function of the flow rates ratio (liquid-to-air) in Figure 3. The flow-pattern map of Baker (1954) as modified by Scott (1963) is also shown in dashed lines in Figure 3. Most of the present data are in the *plug flow* region.

The dimensionless parameters K, F, and T (see eqns 3 through 7) for the present air-water and air-viscoelastic fluids are shown as a function of X in Figures 4 through 6. According to the two-phase generalized-flow-regime map of Taitel and Dukler (1976), the present experimental data may have been in the *intermittent* region (which includes plug flow as well as slug flow).

The frictional pressure drop multipliers, ϕ_l and ϕ_g (see eqns 13, 14, 19 and 20), are shown as a function of X in Figure 7. The present two-phase data analysis revealed that the liquid phase flowing alone in the channel was turbulent and the gas phase flowing alone was laminar. Under these conditions, the value of C in equations 19 and 20 is equal to 10 (Chisholm 1967). The Lockhart -Martinelli graphic relationships for the multipliers (eqs 19 and 20) are also shown as solid and dashed curves (both for C=10) in Figure 7. It may be a coincidence that the eqns 19 and 20 developed for gas-Newtonian liquid two-phase flows predicted the present air-viscoelastic multipliers with reasonable accuracy.

In spite of its deficiencies, most technical calculations are still done using the method of Lockhart-Martinelli. Friedel (1979) proposed improved correlation for two phase pressure drop (in terms of Froude number and Weber number) which yields better results for the case μ_l/μ_g < 1000. Lockhart-Martinelli correlation works better for the case μ_l/μ_g > 1000 and for $\dot{m}_g/A_c$ < 100 kg/(m^2 s)

which the present data satisfy.

The void fraction (ϵ) is shown as a function of Martinelli parameter, X in Figure 8. The present experimental data for air-water and air-viscoelastic fluids are somewhat underpredicted by the famous correlation of Lockhart-Martinelli (1949) proposed for gas-Newtonian two-phase flows.

The experimentally measured turbulent Fanning friction factors for water and for aqueous solutions of Praestol (viscoelastic fluids) are shown in Figure 9. The friction factor data for water, a Newtonian fluid are in agreement with the well-known Blasius equation. The test section (PVC tube) walls are assumed to be "smooth" for engineering purposes. From Figure 9 it can be seen that the friction factor for viscoelastic solutions decreases, at any fixed value of the Reynolds number, with increasing polymer concentration. The friction factors for Praestol solution of 500 wppm concentration have not reached the asymptotic limit. Kwack (1983) proposed an empirical equation (a polynomial in the Weissen-berg number, Ws) for predicting the turbulent friction factor of viscoelastic solutions in straight pipes. Kwack's correlation has very limited applicability since it holds only at two particular values of apparent Reynolds number.

The following equation was developed, based on a regression analysis, for predicting the turbulent friction factor of visco-elastic fluids in straight tubes for $6,000 \leq Re_a \leq 80,000$:

$$f = f_N \, e^{-a\,Ws} \, (1+b\,Ws)/Re_a^c \quad (29)$$

$$f_N = 0.0791 \, Re^{-1/4} \quad (30)$$

where f_N is the turbulent friction factor for Newtonian flows in straight circular pipes (Blasius prediction, Eq 30); a, b, and c are empirical constants. The values of a, b, and c for the present viscoelastic data are -0.0477, 0.004, and 0.0466 respectively. The characteristic times (λ) of these polymer solutions measured by the rheogoniometer are 0.001 s, 0.08 s, and 0.2 s for 100, 200, and 500 wppm concentrations respectively.

Figure 10 makes a graphical comparison of the present experimental turbulent Fanning friction factors for single-phase viscoelastic solutions are compared with the predictions yielded by the Eqn 29. It may be seen that the majority of the present experimental data lie within a region ±20% of the straight line $f_p = f_m$.

The present experimental two-phase friction factors are shown as a function of the two-phase Reynolds number in Figure 11. The f_{TP} and Re_{TP} were calculated using eqns 8 through 12 and 28. As mentioned earlier, ρ_{TP}, μ_{TP} etc can be calculated from the corresponding single-phase quantities, x, and ϵ, using various combinations of mean, and harmonic relationship. The

results reported in this study are based on mean density (using ϵ) and harmonic viscosity (using x). f_{TP} versus Re_{TP} based on other combinations showed little difference compared to the results using the present choice as stated above.

From Figure 11 it can be seen that the two-phase friction factors for air-viscoelastic flows decrease with increasing polymer concentration at any fixed value of Re_{TP}. The minimum concentration for two-phase air-viscoelastic friction factor asymptote is not known. The present data suggest that such an asymptote is not yet reached. The positive displacement pump used in this study posed a restriction on employing further higher concentration of the polymer. The viscosity of these polymer solutions dramatically increases with the concentration (see Figure 2).

Equation 29 (developed to predict single-phase visco-elastic turbulent friction factor) was used to predict the two-phase air-water and air-viscoelastic turbulent friction factor and the results are shown in Figure 12. Eqn 29 predicts the present experimental friction factor data to within ±20%.

The estimated uncertainties, using the root-sum-square method, in the values of friction factors and Reynolds numbers reported in this study are ±5% and ±3% respectively.

The engineering approach uses the *homogeneous* model (simple and at best semi-empirical). In spite of its little *general* applicability to design problems, it is still the most widely accepted in design practice for predicting two-phase pressure drop.

CONCLUSIONS

1) No single flow-regime map considers all the parameters in

classifying the flow-patterns. The present data, based on
the improved Baker's flow-map and Taitel & Dukler's map are
believed to be in the *intermittent* region (more specifically, in the *plug flow*).

2) The well-known Blasius equation predicts turbulent two-phase
friction factor for air-water flows.

3) Air-viscoelastic solutions also exhibit *drag reduction* phenomena
i.e. with increasing polymer concentration, at any fixed
Reynolds number (Re_a), the two-phase turbulent friction factor
monotonically decreases.

4) The new correlation (Eqn 29) may be used to predict two-phase
air-viscoelastic, air-water, and single-phase viscoelastic
fully developed turbulent friction factor in straight pipes of
circular cross section. More data are needed to verify the
validity of this correlation for other polymers.

5) At low pressures (1 to 2 atmospheres), Lockhart-Martinelli
correlation holds for predicting the two-phase pressure drop
for air-water as well as air-viscoelastic turbulent flows in
horizontal pipes. Lockhart-Martinelli correlation underestimates
the void fraction for two-phase flows.

ACKNOWLEDGEMENTS: The author acknowledges the financial support received from Bithell Engineering in completing this work. Dennis Norton took the measurements. Miles Whiting and Terry Snarr contributed by fabricating and instrumenting the apparatus.

REFERENCES

Al-Sheikh, J.N., Saunders, D.E., and Brodkey, R.S. Prediction of flow patterns in horizontal two-phase type flow, *Can. J. Chem. Eng.*, **48**, pp 21-29, 1970.

Baker, O. Simultaneous flow of oil and gas, Oil and Gas J, **53**, p 185, 1954.

Bell, K. J., Taborek, J., and Fenoglio, F. Interpretation of horizontal in-tube condensation heat transfer correlations with a two-phase flow regime map, *Chem. Eng. Prog. Symp. Ser.* **66**, pp 150-163, 1969.

Bird, R. B., Armstrong, R. C., and Hassager, O. Dynamics of

polymeric liquids, vol 1, John Wiley, New York, 1960.

Chisholm, D. A theoretical basis for the Lockhart-Martinelli correlation for two-phase flow, *Int. J Heat Mass Trans*, **10**, pp 1767-1778, 1967.

Cho, Y. I., and Hartnett, J. P. NonNewtonian fluids in circular pipe flow, *Adv in Heat Trans*, **15**, pp 59-141, 1982.

Collier, J. G. Convective Boiling and Condensation, McGraw-Hill, NY 1972.

Dsarasov, Y.I., Kolchugin, B.A., and Liverant, E.I. Experimental study of relation between heat and mass transfer characteristics in evaporating channels, *Proc 5th Int. Heat Trans Conf, Tokyo,* B5.3 4, pp 195-199, 1974.

Fiori, M.P. and Bergles, A.F. A study of boiling water flow regimes at low pressure, M.I.T. Report 3582-40, 1966.

Fisher, S.A., and Yu, S. K. W. Dryout in serpentine evaporators, *Int. J. Multiphase Flow*, **1**, pp 771-791, 1975.

Friedel, L. Improved friction pressure drop calculations for horizontal and vertical two phase pipe flow, *European Two Phase Flow Group Meet., Ispra, Italy,* paper E2, 1979.

Gardner, G. C. Motion of miscible and immiscible fluids in closed horizontal and vertical ducts, *Int. J. Multiphase Flow*, **3**, pp 305-318, 1977.

Hewitt, G. F. *Measurement of*

Two Phase Flow Parameters, Academic Press, New York, NY, 1978.

Kubie, J. The presence of slug flow in horizontal two-phase flow, *Int. J. Multiphase Flow*, **5**, pp 327-339, 1979.

Kutateladze, S.S. Elements of hydrodynamics of gas-liquid system, *Fluid Mech. Sov. Res.*, **1**, pp 29-50, 1972.

Kwack, E.Y. Effect of Weisenberg number on turbulent heat transfer and friction factor of viscoelastic fluids, Ph.D. thesis, University of Illinois, Chicago, 1983.

Lockhart, R. W., and Martinelli, R. C. Proposed correlation of data for isothermal two-phase, two-component flow in pipes, *Chem. Eng. Prog*, **45**, pp 39-48, 1949.

Mandhane, J. M., Gregory, G.A., and Aziz, K.A. A flow pattern map for gas-liquid flow in horizontal pipes, *Int. J. Multiphase Flow*, **1**, pp 537-553, 1974.

McAdams, W. H., Woods, W. K., and Heroman, L.C. Vapourisation inside horizontal tubes, **2**: Benzene-oil mixtures, *Trans. ASME*, **64**, pp 193-200, 1942.

Middleman, S. The flow of high polymers, John Wiley, New York, 1968.

Reimann, J., and John, H. Measurements of the phase distribution in horizontal air-water and steam-water flow, *CNSI Meet. Transient Two-Phase Flow.*, Paris, 1978.

Sakaguchi, T., Akagawa, K., Hamaguchi, H., Imoto, M., and Ishida, S. Flow regime maps for developing steady air-water two-phase flow in horizontal tubes, *Mem. Fac. Eng. Kobe Univ.*, **25**, pp 191-202, 1979.

Schicht, H. H. Flow patterns for adiabatic two-phase flow of water and air within a horizontal tube, *Verfahrenstechnik*, **3**, pp 153-161, 1969.

Scott, D.S. Properties of co-current gas-liquid flow, *Adv in Chem Eng*, **4**, pp 199-277, Academic Press, New York, NY, 1963.

Skelland, A. H. P. NonNewtonian flow and heat transfer, Wiley, New York, 1967.

Taitel, Y., and Dukler, A.E. A model for predicting flow regime transitions in horizontal and near horizontal gas-liquid flow, *AIChEJ*, **22**, pp 47-55, 1976.

Wallis, G. B. One-dimensional two-phase flow, McGraw-Hill, New York, NY 1969.

Wallis, G. B., and Dobson, J.E. The onset of slugging in horizontal stratified air-water flow, *Int. J. Multiphase Flow*, **1**, pp 173-193, 1973.

Wilkinson, W. L. NonNewtonian fluids, Pergamon Press, Oxford, England, 1960.

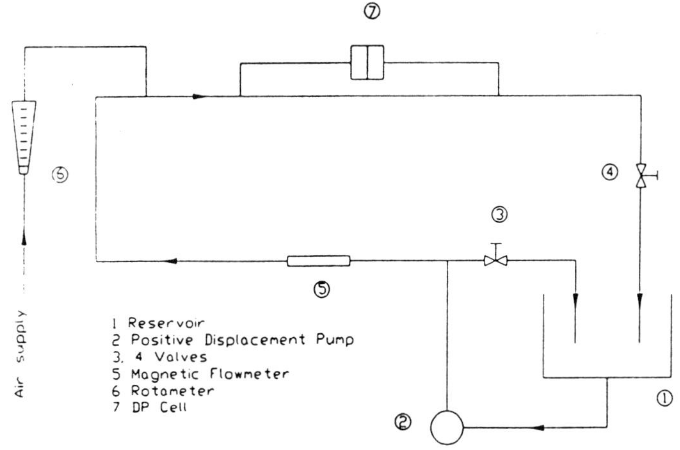

Figure 1. Schematic of flow-loop

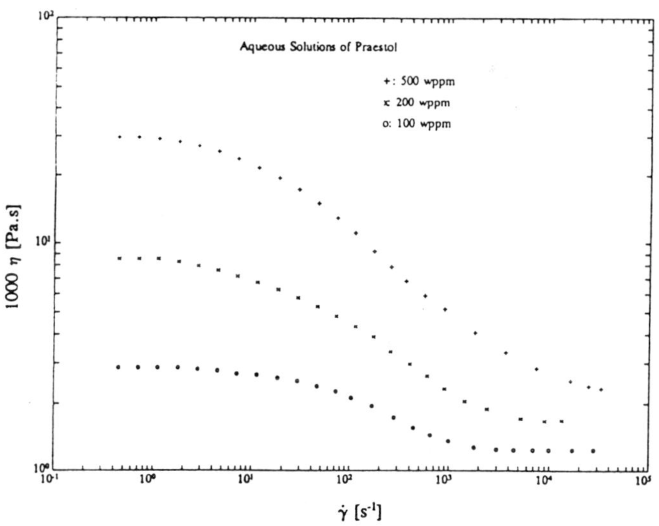

Figure 2. Characteristic Curves of Praestol

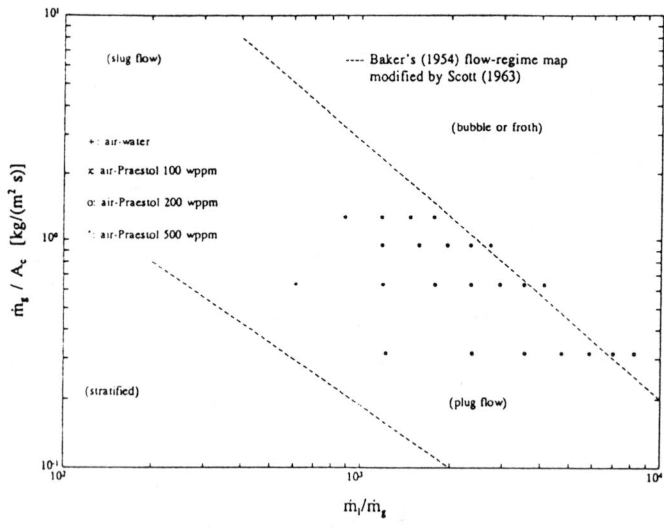

Figure 3. Flow-regime map

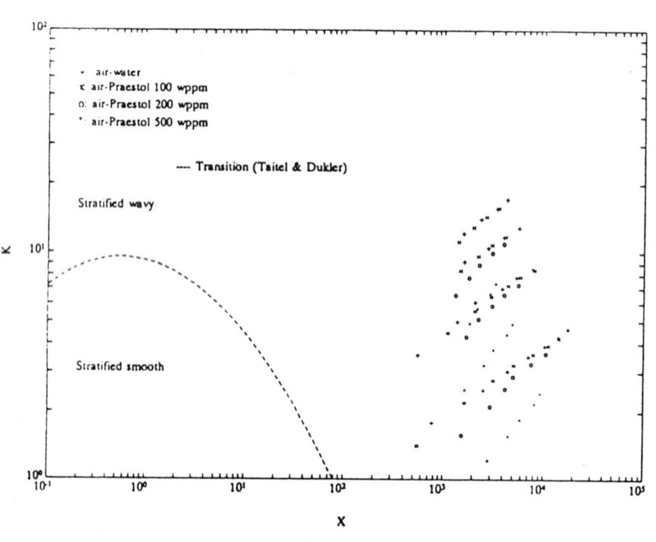

Figure 4. K versus X

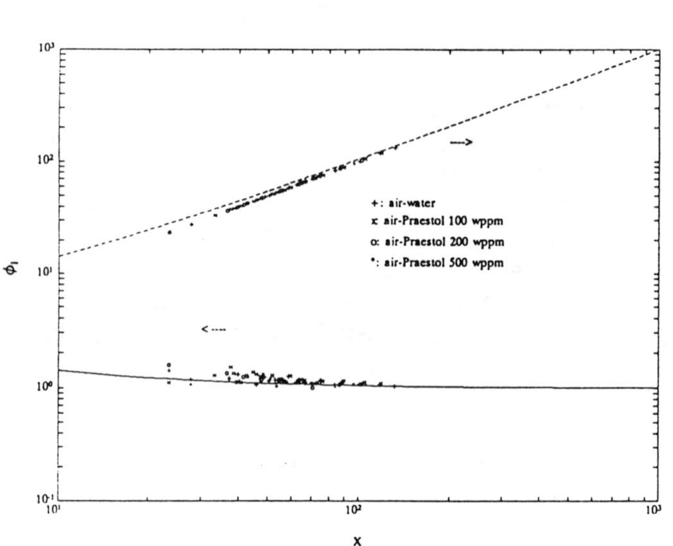

Figure 6. T versus X

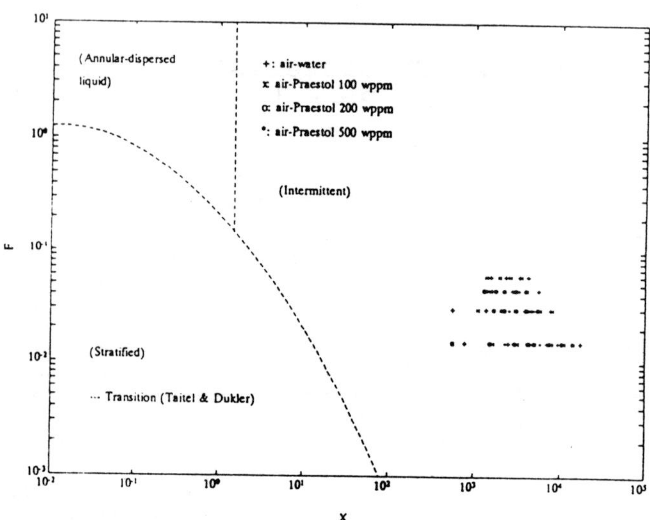

Figure 5. F versus X

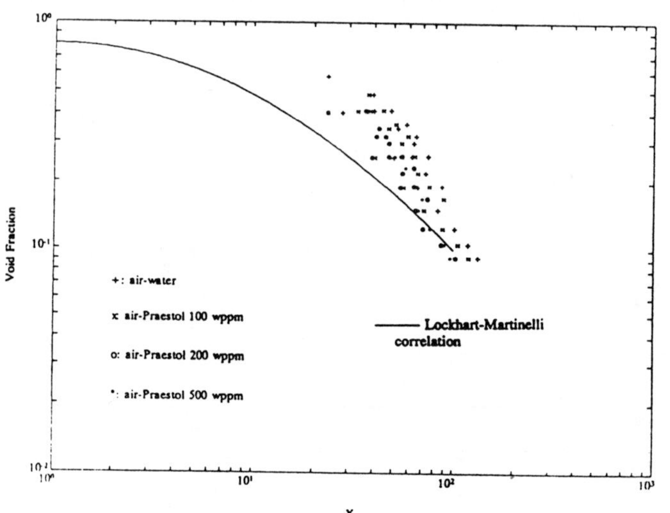

Figure 8. Void fraction versus X

Figure 7. Pressure drop Multipliers

70

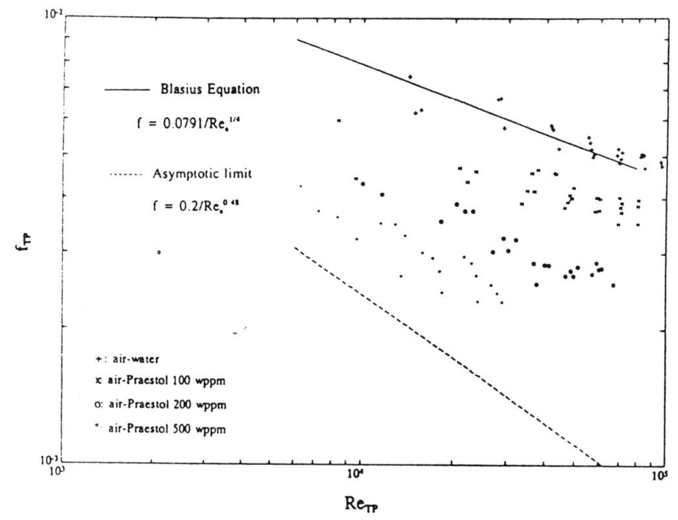

Figure 9. f versus Re$_a$

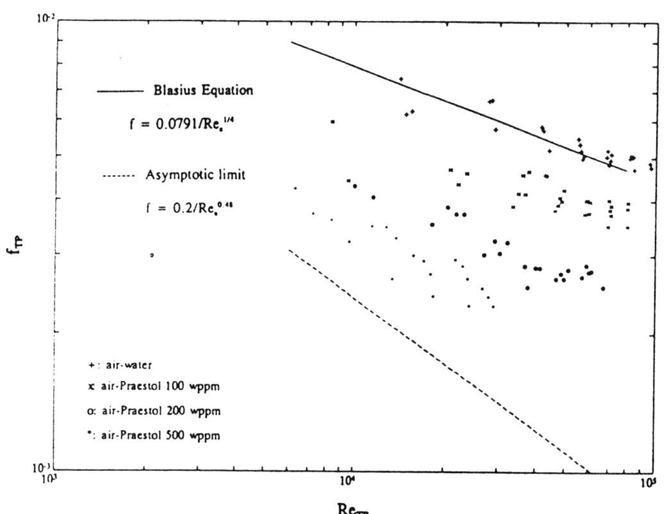

Figure 11. Two-phase f versus Re$_a$

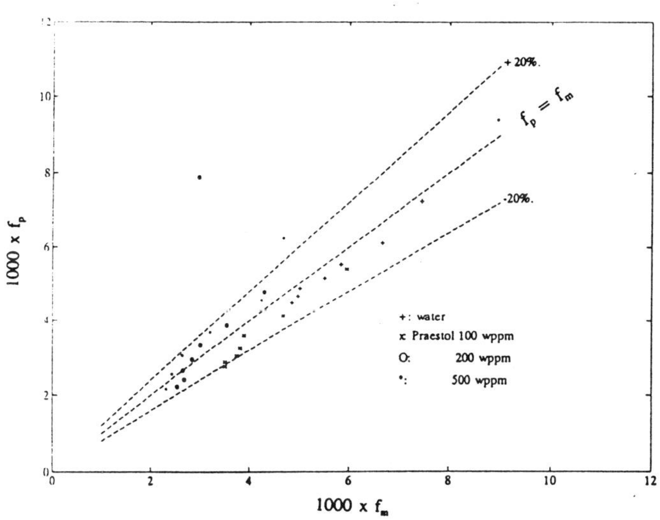

Figure 10. Predicted f versus Measured f

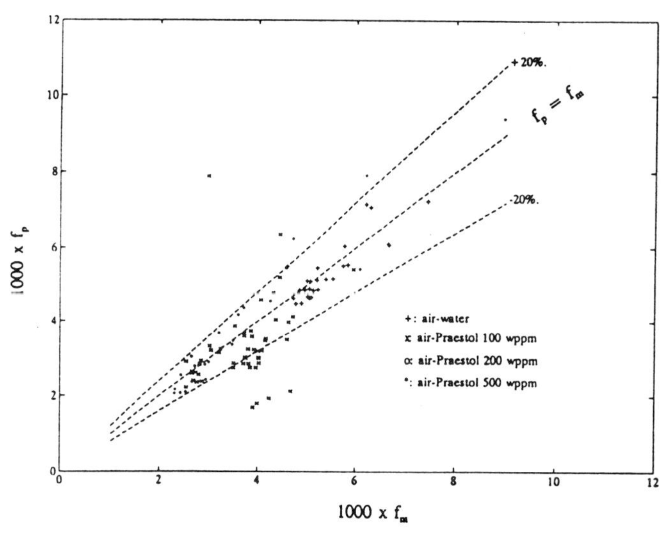

Figure 12. Predicted f versus Measured f

MEASUREMENTS OF CLUSTER-WALL CONTACT TIMES IN A SCALE-MODEL CIRCULATING FLUIDIZED BED

Peter D. Noymer and Leon R. Glicksman
Department of Mechanical Engineering
Massachusetts Institute of Technology
Cambridge, MA 02139 USA

ABSTRACT

One of the mechanisms of heat transfer at the wall of the riser in a circulating fluidized-bed (CFB) is conduction from the clusters of material moving adjacent to the wall; this mechanism is also called particle convection. Relative to each cluster, particle convection is a transient phenomenon, so that the the amount of time that the clusters spend in contact with the wall influences the rate of heat transfer.

The motion of clusters at the wall has been observed and quantified using thermal imaging as a flow-visualization technique. By heating the clusters at the wall and observing their motion with an infrared camera, the average amount of time that a cluster spends at the wall can be determined. The average contact times measured are between 0.35 s and 0.48 s, and some variation with operating conditions in the CFB is observed. These contact times can also be used to calculate bed-to-wall heat-transfer coefficients around 100 W/m^2-K, which is in agreement with other heat-transfer measurements in CFBs. These experiments were conducted in a cold scale-model CFB with a riser that has a 0.159 m square cross-section and is 2.44 m tall.

INTRODUCTION

As circulating fluidized-beds (CFBs) become more widely used as boilers, it becomes beneficial to understand the mechanisms for heat transfer from CFBs. A better understanding of these heat-transfer mechanisms will allow for better designs and more efficient systems. The energy in a CFB is typically exchanged with water tubes that comprise the wall of the riser, so that the mechanisms for heat transfer can be described by the interactions between the bed material, the gas and the wall. In a core-annular flow regime, as seen in the upper portion of a CFB, the flow of bed material near the wall is predominantly downward in agglomerations known as clusters [1, 2, e.g.].

Omitting the radiant heat-transfer contribution (considering it to be additive), the heat transfer from the bed to the wall is typically modeled as a combination of convection from the upward-flowing gas and conduction from the downward-moving clusters, also called "particle convection." The following are aspects of the heat-transfer model: the wall is either covered by gas or clusters, so that gas and particle convection are presumed to act in parallel on the same heat-transfer surface [3]; particle convection consists of clusters exchanging energy via transient conduction with the wall [4], though imperfect contact results in an added thermal resistance [5]; and the gas-convective contribution can be reasonably estimated using correlations for single-phase gas flow [6]. As a result, the total convective heat-transfer at the wall of a CFB is modeled using the following equation:

$$h_{conv} = \left(1 - f\right) \cdot h_{gas} + f \cdot \left(R_{contact} + \sqrt{\frac{\pi\, t}{\left(k\, \rho\, c\,\right)_{cluster}}}\right)^{-1} \quad (1)$$

A number of the heat-transfer parameters in Eq. (1) are functions of the hydrodynamics in a CFB, such as: f, the fraction of the wall covered by clusters; $R_{contact}$, the contact resistance between the cluster and the wall; t, the time of contact between clusters and the wall; and k, ρ and c, the physical properties of the cluster, which are dependent on cluster porosity [7]. Since so much of Eq. (1) is based on hydrodynamic phenomena, the ability to observe, measure and predict these phenomena under various operating conditions would allow for a better understanding and the ability to predict bed-to-wall heat-transfer rates under various operating conditions.

In this work, we are interested in observing and measuring t, the cluster-wall contact time; this is the first time that such measurements have been made directly. Local cluster velocities have been measured previously [1, 8, e.g.] and there have been reports of indirect measurements of the length of contact between clusters and the wall [9, 10]. Typically, one might take the length measurement and divide by the velocity to calculate a contact time. However, the length measurements reported either measure the contact length by cross-correlating heat-transfer rates from probes that are relatively far apart [9] or they are inferred from heat-transfer measurements by using the particle-convective heat-transfer model given in Eq. (1) [10]. These results lack utility for the observation and computation of t, and it becomes apparent that a more direct method is needed for measurement. It was with this in mind that a new technique was developed to measure cluster velocities and contact times between clusters and the wall. This method involves heating the clusters and using their radiant emission to detect their motion at the wall, as discussed in the next section.

DESCRIPTION OF APPARATUS
Scale-Model CFB

All the experiments reported here were conducted in a cold scale-model CFB which was originally built as a 1/4-scale model of a 2-MW$_{th}$ combustor [11]. The riser is 2.44 m high with a square cross-section measuring 0.159 m on each side; the walls are made of clear polycarbonate plastic. There are eleven pressure taps along the riser, so that ten differential-pressure measurements are available for the calculation of the average solid-concentration profile in the riser. The riser has a sharp 90° exit at the top, and the solids are returned to the bottom of the bed via an aerated L-valve. Figure 1 shows a schematic of the scale-model CFB. The bed material is steel powder, with a material density of 6980 kg/m^3 and a mean particle diameter of 69 μm. Using scale-up rules for CFBs [12], it can be shown that this allows for the simulation of a larger CFB combustor operating at atmospheric pressure. Table 1 gives the general specifications of the cold scale-model CFB and the hypothetical full-sized bed being simulated. The simulation of even larger beds is possible with the aforementioned scale-up rules.

Experimental Technique

The basic concept behind the experimental technique is that heating the clusters at the wall alters their emission of thermal radiation and thereby distinguishes them from the rest of the bed. With the appropriate detection device or devices, properties of the flowing clusters can be measured. This has several advantages, among them: it is a non-invasive method for marking the clusters and tracking their motion at the wall, as no instrumentation is required in the flow path; it does not require the addition of tracer particles to the flow to enhance flow visualization, which is beneficial since such particles might affect the hydrodynamics in the riser; and the simple method of marking the clusters with heat does not require the bed material to be any special material itself (such as a photo-luminescent substance) which allows for the selection of a material which will yield similitude with any desired full-sized

configuration. As we are interested in observing the hydrodynamic parameters governing particle convection, using the thermal radiation from the clusters is an appropriate method for observation since the thermal marking method is exactly analogous to particle-convective heat transfer and clusters that are not at the wall are not marked.

This technique requires a test section containing three items: a heater for the clusters; a system capable of detecting infrared radiation; and a wall that is transparent to infrared radiation. In these experiments, the test section is located in the upper portion of the CFB, from 0.92 m above the distributor to the top of the riser (roughly the upper 60% of the bed). The test section is a plane wall, simply replacing one of the walls in that portion of the square bed. The top of the heating section is about 0.48 m from the top of the bed, and it extends 0.15 m down from there. The lower 0.89 m of the test section (as well as the upper 0.48 m) contains the infrared-transparent wall for viewing the motion of the clusters. For the most part, the viewing is done below the heated plate, since the clusters move downward. However, the section above the heated plate is also transparent to infrared radiation so that any upward motion of clusters can be observed (the elevation of the infrared camera can be varied). Figure 1 also shows the location of the test section in the CFB.

The heating section must be flush with the wall of the CFB so that the flow of the solids and the gas remains undisturbed, since small roughness elements on the wall have been shown to disrupt the flow of clusters at the wall [6]. The heating section consists of a resistance heater attached to a thin aluminum plate, where the aluminum plate is the wall of the CFB in that section. The aluminum plate is 0.15 m square and 0.5 mm thick, and the power output of the heater is 190 W. It can be shown that this level of power imparts roughly a 20 K temperature rise to the clusters that contact the plate, and that this temperature rise is sufficient to heat the clusters so that they can be seen in the entire test section. As a cluster travels at the wall after heating, its radiant emission is detectable. When the cluster moves away from the wall its radiant signal will be obstructed by the new material at the wall and it can no longer be detected; however, we are no longer interested in its motion as it is no longer contributing to heat transfer at the wall. This technique allows us to view the history of the clusters which have passed the heater, and it distinguishes them from those clusters that arrive at the wall below the heater.

By heating a cluster at the wall, its temperature is raised above the ambient level in the bed. When running a cold scale-model CFB, these temperatures are on the order of 300 K, so that most of the thermal radiation emitted is in wavelengths around 10 μm. Because of this, special infrared detectors or infrared cameras are required to observe the flow of the marked clusters. Also, most types of glass and plastic materials used for walls of scale-model CFBs absorb all energy at wavelengths above roughly 2 μm. Therefore, a special wall is required to allow for the transmission of the radiant signal from the solid material at the wall. The viewing wall is made of low-density polyethylene (LDPE), also 0.15 m wide and 0.5 mm thick. LDPE was selected as the wall material for its relatively low attenuation of long-wavelength infrared radiation; the 0.5 mm-thick LDPE is roughly 40-50% transparent to radiation at wavelengths of about 10 μm. Since measuring the temperature of the clusters is not our primary concern, we need not be worried that some of the signal is being attenuated. We merely need to be able to detect the radiant signal to indicate the presence of a cluster for these experiments.

A smooth transition between the heating and viewing sections is ensured by making the aluminum plate for the heater and the LDPE wall the same thickness (0.5 mm). Any discontinuity where the two materials meet is filled in with a putty which was sanded down to the same level as the aluminum and LDPE surfaces. A grid-shaped piece of plastic supports the LDPE sheets, where the grid size of 1 cm x 1 cm allows for nearly unobstructed viewing. With some applied tension in the sheet, the grid minimizes the distortion of the LDPE that results from the slight pressurization in the bed. It is unlikely that distortion is eliminated completely; in fact, we estimate that the deflection of the LDPE in the grid is on the order of tens of microns. We recognize that these perturbations on the wall surface are one possible source of error in these experiments. The LDPE sheet is continuous, but there are a discrete number of viewing regions, dictated by the location of the brackets that hold the wall to the rest of the riser. Figure 2 shows the details, from the front and the side, of the construction of the test section.

The infrared detection system consists of an infrared camera and a VCR to record the signal. The camera is the Model 600 Infrared Imaging Radiometer from Inframetrics, Inc., which has a mercury/cadmium/telluride (HgCdTe) detector that can be set to detect radiation in the 8-12 μm portion of the spectrum. The output is a video signal at the standard video rate of 30 Hz; this is wired into a conventional VCR. Since the camera is mounted on a tripod, it can be aligned with any of the viewing windows and it can be set back at such a distance to capture the entire window in its field of view. With conventional video output, the data is reduced by analyzing the videotape in slow motion.

Experimental Operating Conditions

The scale-model CFB was run at a total of eight different operating conditions for these experiments. The gas velocities and solid recycle rates were varied in order to understand the effect of operating conditions on the velocities and contact times being measured. Table 2 lists the different operating conditions, along with the average local solid concentrations in the test section.

RESULTS
Overview

To measure the contact times of clusters at the wall of the CFB, the infrared camera was kept focused on each viewing port (see again Fig. 2) for a certain period of time. By analyzing the resulting videotape in slow motion, the average number of clusters passing by a particular viewing port in a given period of time were measured; we call this the "passing frequency" of clusters, and there is obviously some subjectivity in defining a cluster. In analyzing the videotape, the top and bottom halves of the first viewing port were analyzed separately, allowing for slightly greater resolution of the passing frequencies.

The viewing ports are located at fixed distances below the heated plate; therefore, the passing frequency in each viewing port actually represents the passing frequency as a function of the distance traveled after heating. Since the average velocity of the clusters in each viewing port can be measured for each operating condition, the passing frequency is also known as a function of the travel time below the heated plate. If normalized, the passing frequency can be thought of as a probability function for observing a cluster a certain time after heating. With that, the rate of change of the passing frequency, or the "shedding function," is the probability function for a cluster disappearing from the wall a certain time after heating. Restated, this is the probability that a cluster remains in contact with the wall a certain time after heating, which is precisely the information we seek.

With the infrared camera, only the clusters that have been heated appear in the viewing ports; this allows us to identify clusters as having existed at the wall at or before the end of the heating section. Any cluster deposited at the wall after the heating section will not appear in the camera. This is an important distinction, since fixing the bottom of the heating section as a reference point for the observed clusters is critical to the measurement of the cluster-wall contact times. Furthermore, when a cluster disappears from view, or when a change in passing frequency is measured, it can be assumed that the cluster was shed from the wall. The shedding of a cluster ends its contact with the wall and its contribution to heat transfer at the wall. Since the detection of clusters depends on their temperature, it may be possible that the disappearance of clusters results from cooling; as discussed earlier, it can be shown that the heater has enough power to sufficiently raise the temperature of almost all of the clusters such that they can be observed, or such that they have not cooled down, by the end of the entire test section. This ensures that the measured changes in the passing frequency are due to clusters moving away from the wall, and not from clusters cooling down.

Cluster-wall contact times – measurements

The cluster-wall contact times were measured under the operating conditions listed in Table 2. Again, contact times are derived from the rate of change of the measured passing frequencies of clusters. Data for passing frequencies were obtained by sampling 60 seconds of video data at 30 Hz in each viewing port. In analyzing the videotape, each 60-second sample was subdivided into about 10 sets; the average passing frequency for each of these 5-6 second sets was measured, and then that group was averaged to yield an overall average. Subdividing the 60-second sample into a number of sets allows for the estimation of statistical confidence limits on the group average; this was typically found to be about ±10% for 90% confidence limits. Furthermore, the visual analysis of a sample was determined to be repeatable within about 5%, so that the total uncertainty on a given average passing frequency is about ±15%. By using the velocities of the clusters to convert the abscissa from distance to time, we introduce the uncertainty error associated with the velocity measurements (roughly ±30%). Using a root-mean-square method for combining uncertainties [13], the total uncertainty in the measurement of the contact times is about ±35%.

Figure 3 shows the passing frequency function for one of the operating conditions, the shape of which is typical of all of the operating conditions. Note that the type of curve that fits the passing frequencies best is an exponential function. Since the shedding function is the negative of the derivative of the passing frequency, the shedding function will also have an exponential shape with the same time constant. If the shedding function is normalized such that the area under its curve is unity, then the shedding function is equivalent to the probability density function for a cluster traveling a certain time after heating, as discussed previously. This normalized shedding function is also shown with the measurements in Figure 3. An exponential probability function, like the normalized shedding function that we can calculate from fitting the measurements of passing frequencies, is defined by:

$$m(\tau) = \frac{1}{\tau_{avg}} \cdot exp\left(-\frac{\tau}{\tau_{avg}}\right) \tag{2}$$

where τ is the random variable and the average value is simply τ_{avg}. Since exponential functions fit the frequency and shedding functions best, the average contact times after heating for each of the cases are simply the decay constants of the fitted frequency functions; these values are contained in Table 2.

A probabilistic analysis shows that the total cluster-wall contact time is twice the measured time of contact after heating [14]. In other words, $t_{avg} = 2 \cdot \tau_{avg}$; these results are also given in Table 2 and they are plotted in Figures 4(a)-(c). These figures show how the total cluster-wall contact times vary with the superficial velocity of the gas, the solid recycle rate, and the average local cross-sectional solid concentration. For all three parameters, a slight downward trend is apparent, but it is difficult to confirm the existence of these trends as they fall within the range of experimental uncertainty. At this time, these are the only measurements of cluster-wall contact times, so it is impossible to gather other data for comparison.

One way to evaluate the soundness of these contact-time measurements is to use them in conjunction with the model presented in Eq. (1) to calculate bed-to-wall heat-transfer coefficients. This model requires knowledge of several other hydrodynamic parameters, but estimates for these can be made based on previous research [6]: $f \approx 0.50$, $h_{gas} \approx 12$ W/m^2-K, $R_{contact} \approx 0.0024$ m^2-K/W, and $\sqrt{(k\rho c)_{cluster}} \approx 100$ m^2-K/W-s$^{1/2}$. For the contact times measured, then, we can calculate heat-transfer coefficients around 100-125 W/m^2-K. These calculations are in agreement with measurements of heat-transfer coefficients in other cold scale-model CFBs, generally found to be about 50-250 W/m^2-K [15].

CONCLUSIONS

In a cold scale-model CFB, we have used a new experimental technique to measure the contact times of clusters at the wall of a CFB. We found that for a smooth-walled CFB of square cross-section with 69 μm steel powder as the bed material, the cluster-wall contact times are between 0.35 and 0.50 s. Some variation in the data was observed with operating conditions (gas and solid flow rates). Furthermore, these contact times can be used to calculate heat-transfer coefficients, based on the model presented in Eq. (1), that are consistent with typical measurements for cold scale-model CFBs.

ACKNOWLEDGEMENTS

This work was sponsored by the National Science Foundation. The authors are grateful to MIT/Lincoln Laboratories and the Welding Laboratory at MIT for the use of their infrared cameras, and to the Fluid Mechanics Laboratory at MIT for the use of their VCR.

NOMENCLATURE

c	specific heat capacity (J/kg-K)
d_p	mean particle diameter (m)
D	bed diameter (m)
f	fraction of wall covered by clusters (–)
G_s	solid recirculation rate (kg/m^2-s)
h	heat-transfer coefficient (W/m^2-K)
k	thermal conductivity (W/m-K)
L	riser height (m)
$m(\tau)$	probability distribution of measured contact times (–)
Nu	Nusselt number (–)
P	bed pressure (atm)
R	thermal resistance (m^2-K/W)
t	actual contact time between cluster and wall (s)
T	bed temperature (K)
u_0	gas superficial velocity (m/s)
ε	volumetric void fraction (–)
ρ_g	gas density (kg/m^3)
ρ_s	solid density (kg/m^3)
τ	measured contact time between cluster and wall (s)

REFERENCES

1. Ishii, H., Nakajima, T., and Horio, M., "The Clustering Annular Flow Model of Circulating Fluidized Beds," *Journal of Chemical Engineering of Japan*, **22**, 484-490 (1989).

2. Soong, C.H., Tuzla, K., and Chen, J.C., "Identification of Particle Clusters in Circulating Fluidized Beds," in Circulating Fluidized Bed Technology IV, A. Avidan, ed., 615-620 (1993).

3. Subbarao, D., and Basu, P., "A Model for Heat Transfer in Circulating Fluidized Beds," *International Journal of Heat and Mass Transfer*, **29**, 487-489 (1986).

4. Mickley, H.S., and Fairbanks, D.F., "Mechanism of Heat Transfer to Fluidized Beds," *AIChE Journal*, **1**, 374-384 (1955).

5. Baskakov, A.P., "The Mechanism of Heat Transfer Between a Fluidized Bed and a Surface," *International Chemical Engineering*, **4**, 320-324 (1964).

6. Lints, M.C., and Glicksman, L.R., "Parameters Governing Particle-to-Wall Heat Transfer in a Circulating Fluidized Bed," in Circulating Fluidized Bed Technology IV, A. Avidan, ed., 297-304, (1993).

7. Gelperin, N.I., and Einstein, V.G., "Heat Transfer in Fluidized Beds," in Fluidization, J.F. Davidson and D. Harrison, eds., Academic Press, 471-540 (1971).

8. Rhodes, M., Mineo, H., and Hirama, T., "Particle Motion at the Wall of a Circulating Fluidized Bed," *Powder Technology*, **70**, 207-214 (1992).

9. Wu, R.L., Lim, C.J., Grace, J.R., and Brereton, C.M.H., "Instantaneous Local Heat Transfer and Hydrodynamics in a Circulating Fluidized Bed," *International Journal of Heat and Mass Transfer*, **34**, 2019-2027 (1991).

10. Fang, Z.H., Grace, J.R., and Lim, C.J., "Local Particle-Convective Heat Transfer Along Surfaces in Circulating Fluidized Beds," *International Journal of Heat and Mass Transfer*, **38**, 1217-1224 (1995).

11. Westphalen, D., "Scaling and Lateral Solid Mixing in Circulating Fluidized Beds," Ph.D. Thesis, Massachusetts Institute of Technology, Cambridge, MA (1993).

12. Glicksman, L.R., Hyre, M., and Woloshun, K., "Simplified Scaling Relationships for Fluidized Beds," *Powder Technology*, **77**, 177-199 (1993).

13. Beckwith, T.G., Marangoni, R.D., and Lienhard, J.H. V, Mechanical Measurements, 5th Ed., Chap. 3, Addison-Wesley, Reading, MA (1993).

14. Panta, S.B., "A Probabilistic Analysis for the Interpretation of Flow-Visualization Data at the Wall of a Circulating Fluidized Bed," B.S. Thesis, Massachusetts Institute of Technology, Cambridge, MA (1996).

15. Basu, P., and Nag, P.K., "Heat Transfer to Walls of a Circulating Fluidized Bed Furnace," *Chemical Engineering Science*, **51**, 1-26 (1996).

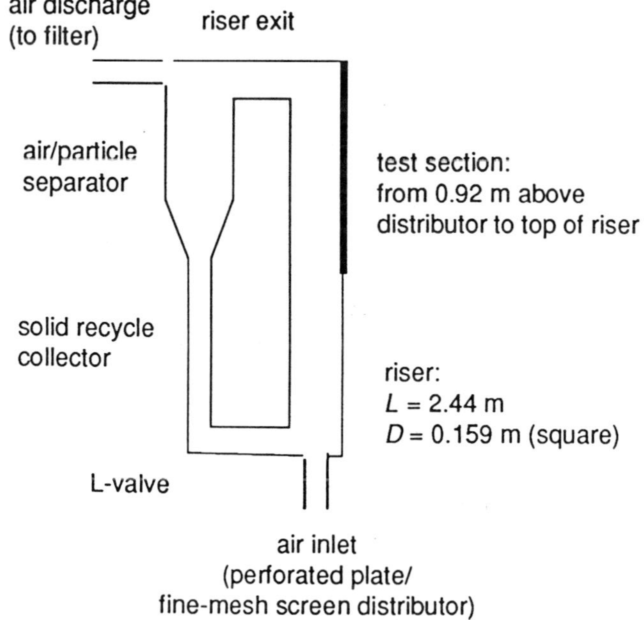

FIGURE 1: Schematic for scale-model CFB

Parameter	scale-model	4:1 scale-up
L (m)	2.44	9.76
D (m)	0.159	0.636
x-section shape	square	square
T (K)	300	1100
P (atm)	1.0	1.0
ρ_g (kg/m^3)	1.2	0.33
ρ_s (kg/m^3)	6980	$\approx$ 2500
d_p (μm)	69	$\approx$ 230
typical u_0 (m/s)	3	6
typical G_s (kg/m^2-s)	25	18

TABLE 1: Specifications for scale-model and hypothetical full-sized CFBs

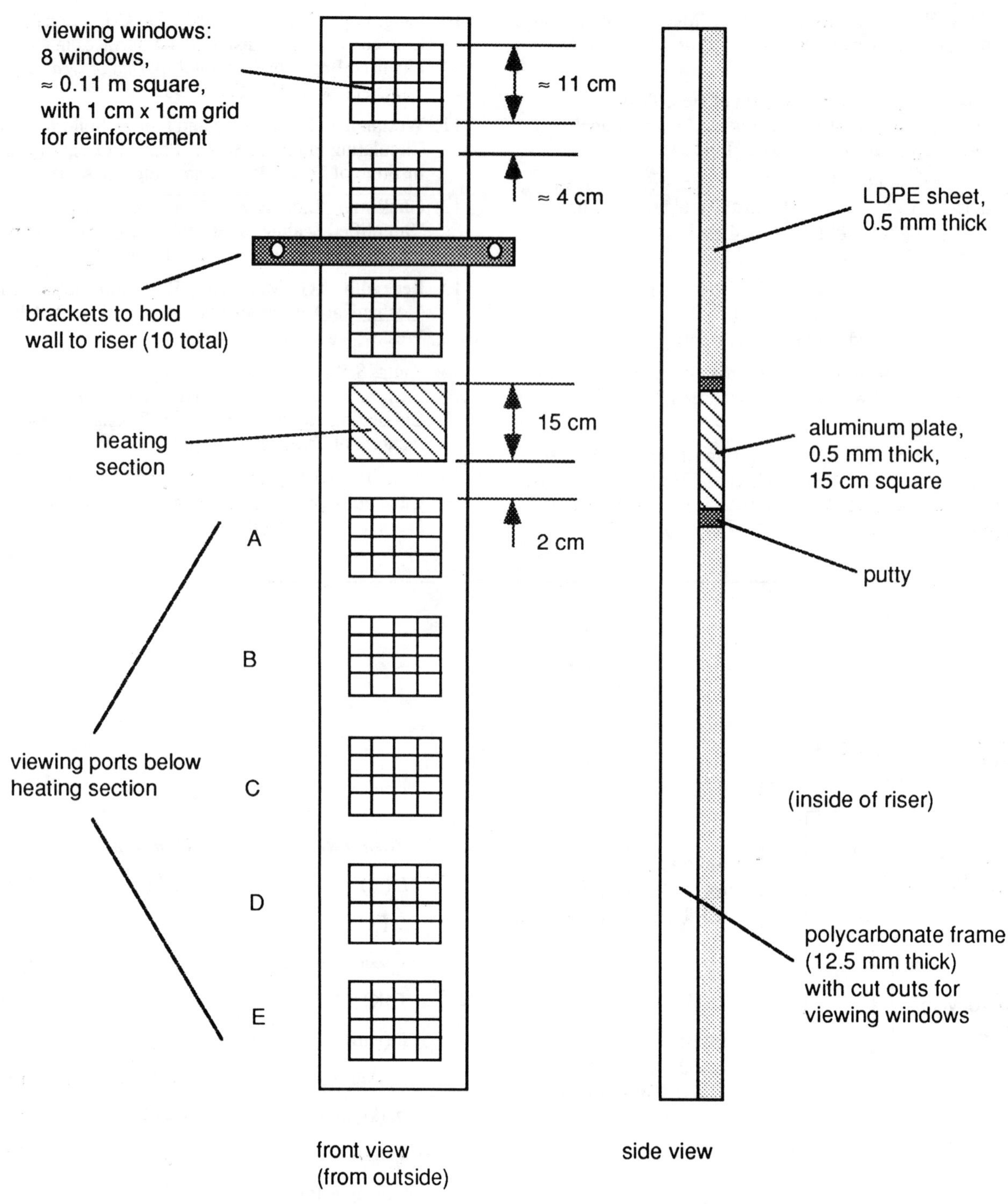

FIGURE 2: **Construction details for the wall in the test section**

Case number	u_o (m/s)	G_s (kg/m²-s)	avg. local solid conc.	avg. time below heating section (s)	avg. total contact time (s)
1	2.3	9.3	0.25%	0.24	0.49
2	2.8	11	0.21%	0.19	0.37
3	2.8	19	0.46%	0.22	0.45
4	2.7	30	0.56%	0.19	0.37
5	3.3	18	0.24%	0.19	0.37
6	3.3	28	0.42%	0.21	0.42
7	3.6	25	0.28%	0.19	0.38
8	3.6	34	0.49%	0.17	0.33

TABLE 2: Experimental operating conditions and results

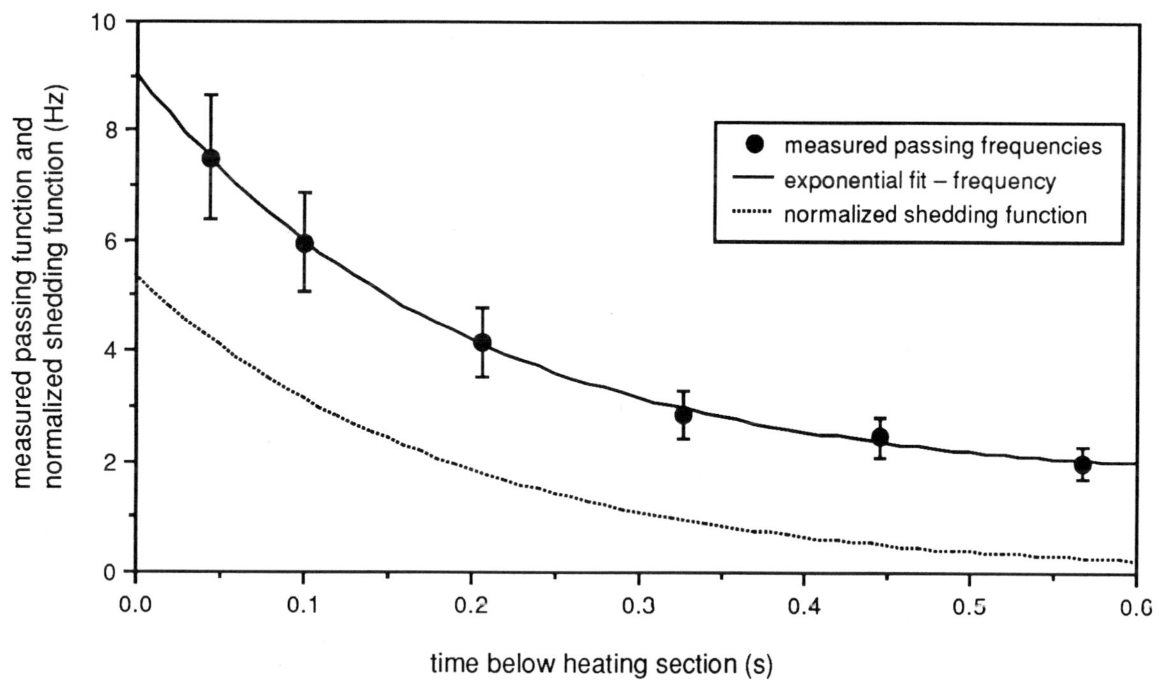

FIGURE 3: Typical measured passing frequencies of clusters and exponential fit (from Case 5)

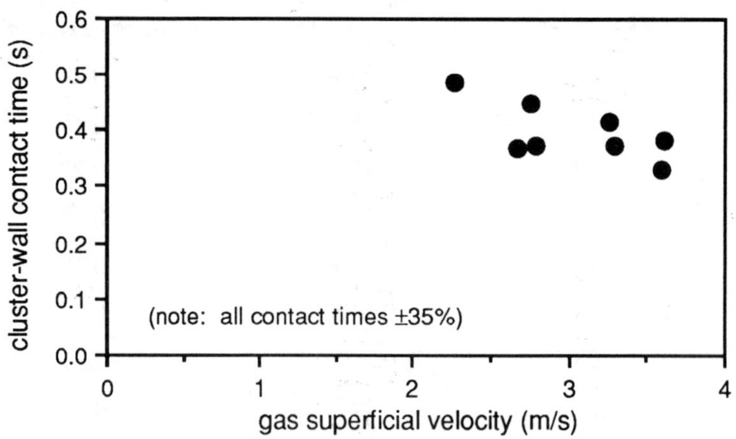

FIGURE 4(a): Cluster-wall contact time vs. gas velocity

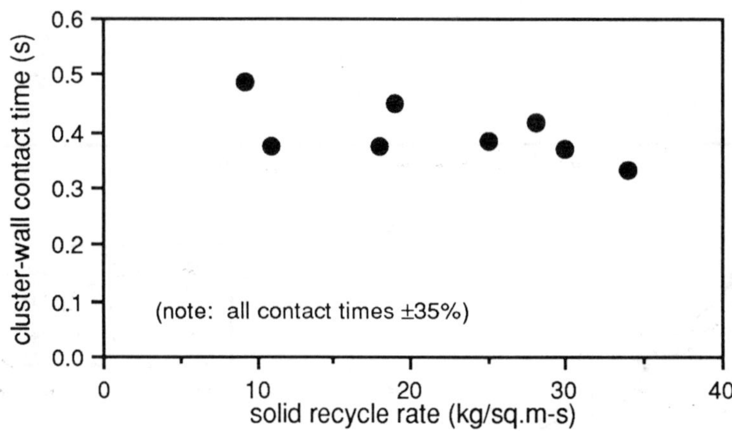

FIGURE 4(b): Cluster-wall contact time vs. solid recycle rate

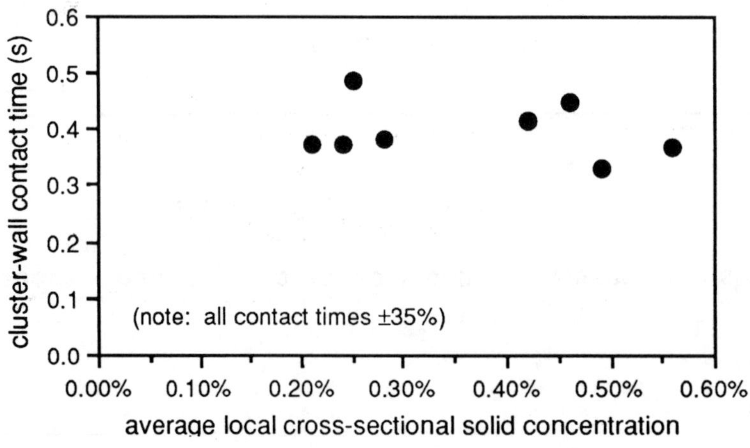

FIGURE 4(c): Cluster-wall contact time vs. solid concentration

STRATIFICATION IN NUCLEAR REACTOR COMPONENTS

J. A. Randorf
Y. A. Hassan

Department of Nuclear Engineering
Texas A&M University
College Station, Tx-77843-3133

ABSTRACT

Several experiments have recorded loop stratification at a recent test performed at the Rig of Safety Assessment (ROSA) project at the Japanese Atomic Energy Research Institute's Large Scale Test. This facilty (LSTF) is used for for the advanced nuclear reactor AP600 simulation. Prediction of stratification in reactor coolant loops is of interest during accident conditions. The present study attempts to develop the criteria for stratification in the coolant loops based on data from the LSTF experiments. The first criteria was developed for addressing injection-induced stratification. The second criteria applies to downcomer/cold leg junction stratification. Both criteria provided predictions consistent with LSTF data.

NOMENCLATURE

Fr = Froude Number
Re = Reynolds Number
V = velocity (m/s)
g = gravitational acceleration(m/s^2)
D = diameter of pipe(m)
L = stratification length (m)
Q = Volumetric flow rate(m^3/s)
ρ = Density (kg/m^3)
β = Volumetric Expansion Coefficient
c = 0.012 kg$^{1/2}$ m$^{-7/2}$ s$^{5/2}$

INTRODUCTION

In a nuclear power plant, the injection of emergency core cooling system (ECCS) water can cause buoyancy induced stratification in the injection loops. Accidents of this type are studied extensively, an example being the ROSA-V large scale test facility (LSTF). A series of experiments were conducted to verify the design of the Westinghouse designed Advanced Passive 600 Mwe (AP600) reactor at the LSTF.

Using ROSA data, the applicability of previously developed stratification criteria was studied. This was a Froude number based criterion developed by Theofanous (1984). This criterion had provided reasonable agreement with the data.

A second criterion based on Reynolds number and Froude number developed by Opanasenko (1995) was also tested against this data. This criterion applies to the downcomer / cold leg junction. Thermal and flow transients promote back flow into the cold leg, producing stratification against the incoming flow. This criterion accounts for the viscous effects.

STRATIFICATION CRITERIA

The Froude number can be considered as the ratio of the inertial forces over the gravitational forces. The following form of the densimetric modified Froude number is used in this study:

$$Fr = \frac{V}{\sqrt{gD\dfrac{\Delta\rho}{\rho}}} \qquad (1)$$

The density difference is between the injected coolant (heavy liquid) and the primary coolant loop flow (lighter liquid).

The change in density can be approximately shown as the product of the fluid temperature difference and the average temperature expansion coefficient. Theofanus et al. (1984) showed that the ratio of the two extremes (perfectly mixed and perfectly stratified) can be represented as:

$$\frac{\rho_{mix} - \rho_{loop}}{\rho_{loop}} = \beta_{mix}(T_{loop} - T_{mix}) \qquad (2)$$

$$\frac{\rho_{prhr} - \rho_{loop}}{\rho_{prhr}} = \beta_{prhr}(T_{loop} - T_{prhr}) \qquad (3)$$

This results in the Froude number based on the density differences between e.g; the loop and the Passive Residual Heat Removal (PRHR) flows as:

$$Fr_{prhr,loop} = \left[1 + \frac{Q_{loop}}{Q_{prhr}}\right]^{-\frac{3}{2}} \sqrt{\frac{\beta_{mix}}{\beta_{prhr}}} \qquad (4)$$

By representing the effects of non-linear fluid expansion within the exponent of the first product, thereby removing the square root term, the above relation simplifies to:

$$Fr_{prhr,loop} = \left[1 + \frac{Q_{loop}}{Q_{prhr}}\right]^{-\frac{7}{5}} \qquad (5)$$

Loop flows that have a Froude number less than those given by equation (5) are considered stratified, those greater, well-mixed. This relationship was developed for injection systems that insert fluid above horizontal channels.

The other criterion that was implemented to determine the onset of stratification is based on the Reynolds and Froude numbers parameters. The Reynolds number is defined as the ratio of the inertial forces to the viscous forces and given by:

$$Re = \frac{VD\rho}{\mu} \qquad (6)$$

Mixing is enhanced with highly turbulent flows (high Reynolds numbers). This suggests that the Reynolds number could be used directly in predicting stratification.

Some studies (Opanasenko, 1995) correlated the ratio of stratification length over channel width in a horizontal pipe. Similar situation as the horizontal cold leg/ downcomer annulus. The relation is given as:

$$c = \frac{L/D}{Re^{1/2}Fr^{-3}} \qquad (7)$$

In our analysis we have assumed that stratification begins when L/D = 0. The correlation is then rewritten as:

$$\ln(Fr) = \frac{1}{3}\ln(c) + \frac{1}{6}\ln(Re) \qquad (8)$$

The value of c is relatively constant regardless of the system specific geometry and given a value of about 0.012 for pipe diameter of 45 mm and 0.015 for diameter of 75 mm.

ANALYSIS

The analysis to predict the onset of stratification was performed utilizing the test for 1-inch cold leg break of ROSA/AP600 facility. Schematic of the test facility is shown in Fig. 1. During a portion of the scenario for the 1-inch break test, thermal stratification in cold leg A (CL-A) and cold leg B (CL-B) was observed.

Loop A Stratification

The ROSA data (Shaw et al., 1994) indicated stratification in the loop-A. In the ROSA facility, the

PRHR injection location is downstream of steam generator outlet. The cold water, from the PRHR injection, flowed into the bottom of the cold leg while, the warmer liquid of the cold leg flowed at the top. To compute the Froude number, the total loop flow upstream of the PRHR injection point was required. Additionally, the RHR injection rate and the loop and the injection temperatures are required. The maximum instrument error was 3.38%. In loop-A, the onset of stratification is at 375 seconds into the transient. The abscissa in Fig. 2 presents the ratio of total flow to PRHR injection rate. The left ordinate is the Froude number values, while, the right ordinate shows the maximum difference between the hottest and the coldest thermocouple temperatures in the cold leg. The time when the Froude number crosses the criterion line (estimated from equation 5) at the specified volumetric flow correspond well with the onset of stratification. The slope of the criterion line is -7/5.

Loop-B Stratification

In Loop-B, the stratification started at the connection of the cold leg with the downcomer. Relatively, cold water of the downcomer intruded into the bottom of the cold leg. In this case the logarithmic equation (8) is used to determine the stratification criterion. The data is plotted in Fig. 3. The abscissa is the logarithm of the Reynolds number and the primary ordinate is the logarithm of the cold leg Froude number. The right ordinate presents the temperature difference between the cold and hot liquids. The criterion line is the solid line which is the natural log of the Reynolds number plus the natural log of the constant C. The slope is 1/6. A reasonable estimate of the onset of the stratification is obtained.

CONCLUSIONS

Two methods for predicting stratification in reactor coolant loops are presented. The first uses the ratio of injected to loop flow as a prediction criterion whereas the second uses the Reynolds number and the Froude number as the stratification criterion. Both approaches produce reasonable results. As loop flow increases or injection flow drops, the mixing tends to increase, thus providing conditions for a smaller Froude number values less than 0.2.

REFERENCES

Theofanus, T. G. et al., " Decay of Buoyancy Driven Stratified Layers with Applications to Pressurized Thermal Shock", NUREG/CR-3700, May 1984.

Opanasenko, A., " Thermal stratification of Coolant in Nuclear Power Installation Structural Components", The State Scientific Center of Russian Federation, The Institute of Physics and Power Engineering, Bondarenko, Russia, December 1995

Shaw R. A. , Yonomoto, T. and Kukita Y., " Quick Look Report for ROSA/AP600 Experiment AP-CL-03". JAERI Report M 06-249, 1994.

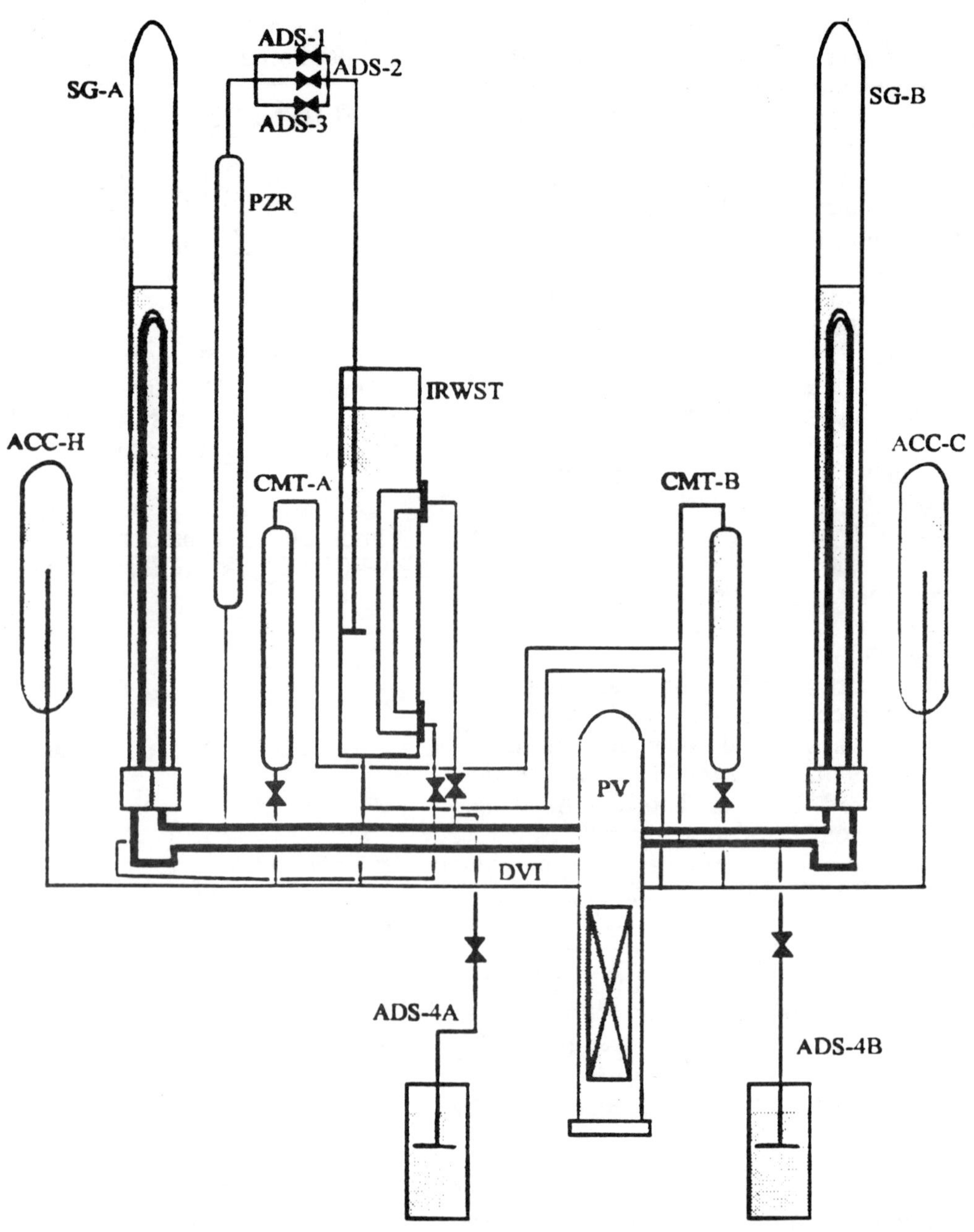

Fig. 1 Schematic of ROSA/AP600 Large Scale Test Facility

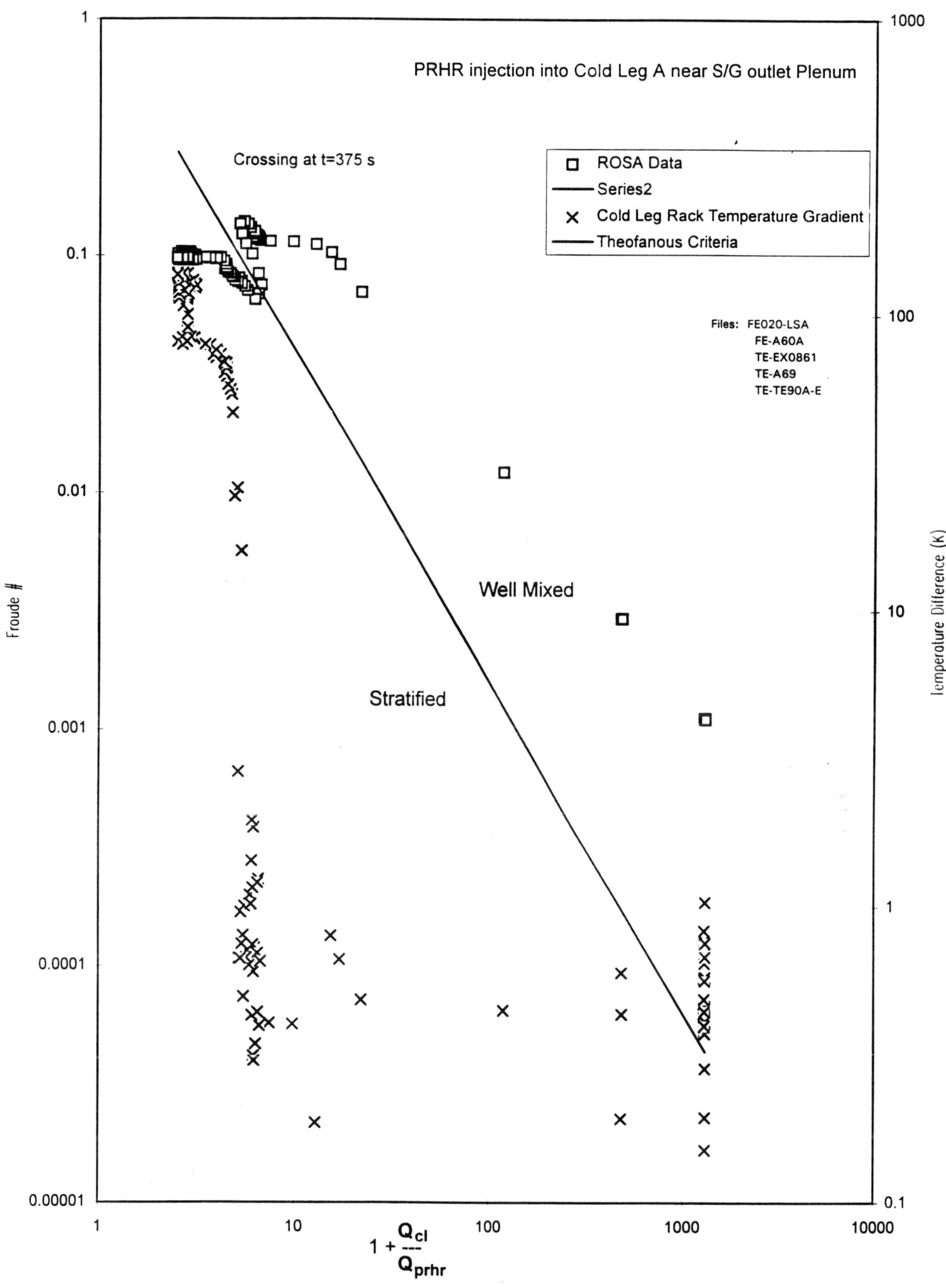

Fig. 2 Stratification Criterion in Loop A

85

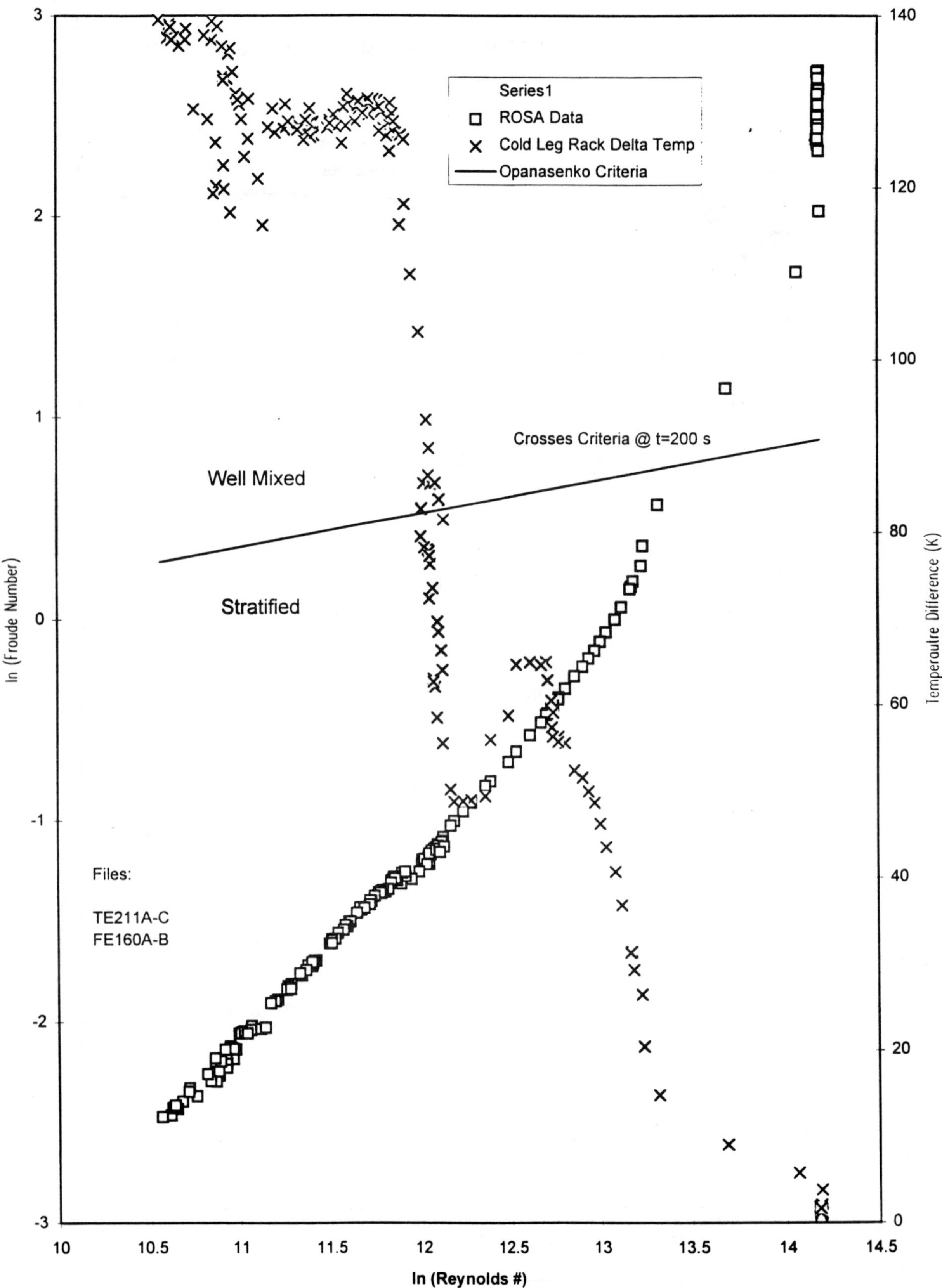

Fig. 3 Stratification Criterion in Loop B

86

AUTHOR INDEX

NE-Vol. 19
Thermal Science of Advanced Steam Generator/Heat Exchangers